XII Serbian-Bulgarian Astronomical Conference (SB 2020)

Publications of the Astronomical Society "Rudjer Boskovic"
No. 20

Sokobanja, Serbia
25 – 29 September 2020

Editors:

Luka C. Popovic
Vladimir A. Sreckovic
Milan S. Dimitrijevic
Andelka Kovacevic

ISBN: 978-1-7138-2909-6

Printed from e-media with permission by:

Curran Associates, Inc.
57 Morehouse Lane
Red Hook, NY 12571

Some format issues inherent in the e-media version may also appear in this print version.

Printed with permission by Curran Associates, Inc. (2021)

For permission requests, please contact Serbian Astronomical Society
at the address below.

Serbian Astronomical Society
Volgina 7
11000 Belgrade
Serbia

Phone: +381-11-3089-072

mdimitrijevic@aob.bg.ac.rs

Additional copies of this publication are available from:

Curran Associates, Inc.
57 Morehouse Lane
Red Hook, NY 12571 USA
Phone: 845-758-0400
Fax: 845-758-2633
Email: curran@proceedings.com
Web: www.proceedings.com

ПУБЛИКАЦИЈЕ АСТРОНОМСКОГ ДРУШТВА "РУЂЕР БОШКОВИЋ"
PUBLICATIONS OF THE ASTRONOMICAL SOCIETY "RUDJER BOŠKOVIĆ"

Св. 20 No. 20

PROCEEDINGS OF THE XII SERBIAN-BULGARIAN ASTRONOMICAL CONFERENCE

Sokobanja, Serbia, September 25-29, 2020

Eds. Luka Č. Popović, Vladimir A. Srećković,

Milan S. Dimitrijević and Anđelka Kovačević

БЕОГРАД
2020

PUBL. ASTRON. SOC. "RUDJER BOŠKOVIĆ" No. 20, 1-192 BELGRADE, DECEMBER 2020

SCIENTIFIC COMMITTEE

Luka Č. Popović (Co-chairman)
Vladimir A. Srećković (Co-chairman)
Ognyan Kounchev (Co-chairman)

Milan S. Dimitrijević (Co-vice chairman)
Milcho K. Tsvetkov (Co-vice chairman)

Robert Beuc (Croatia)
Svetlana Boeva (Bulgaria)
Momchil Dechev (Bulgaria)
Dragana Ilić (Serbia)
Andjelka Kovačević (Serbia)
Jelena Kovačević Dojčinović (Serbia)
Maša Lakićević (Serbia)
Petko Nedialkov (Bulgaria)
Nikola Petrov (Bulgaria)
Branko Predojević (Bosnia and Herzegovina)
Milan Radovanović (Serbia)
Saša Simić (Serbia)
Nikolaj Samus (Russia)
Georgi Simeonov (Bulgaria)
Katya Tsvetkova (Bulgaria)
Dejan Urošević (Sebia)
Jan Vondrák (Czech Republic)

LOCAL ORGANIZING COMMITTEE

Andjelka Kovačević (Chairperson)

Maša Lakićević (Scientific secretary)

Members:
Milan S. Dimitrijević
Jelena Kovačević Dojčinović
Slađana Marčeta Mandić
Saša Simić
Vladimir A. Srećković

ORGANIZER:
Astronomical Observatory Belgrade, Serbia

Co-organizers:
Institute of Mathematics and Informatics - BAS, Sofia, Bulgaria
Institute of Physics Belgrade, Belgrade, Serbia
Faculty of Mathematics and Department of Astronomy, University of Belgrade, Serbia

CONTENTS

Goran Damljanović
GAIA DR3 AND SOME RESULTS OF SERBIAN-BULGARIAN COOPERATION 5

Goran Damljanović, Rumen Bachev, Svetlana Boeva, Georgy Latev, Milan Stojanović, Miljana D. Jovanović, Oliver Vince, Zorica Cvetković, Rade Pavlović and Gabrijela Marković
SERBIAN-BULGARIAN OBSERVATIONS OF GAIA ALERTS (GAIA-FUN-TO) DURING 2019 15

Miljana D. Jovanović, Goran Damljanović, Zorica Cvetković, Rade Pavlović and Milan Stojanović
COLOR VARIABILITY OF SOME QUASARS IMPORTANT TO THE ICRF – GAIA CRF LINK 23

Jelena Kovačević-Dojčinović, Ivan Dojčinović, Maša Lakićević and Luka Č. Popović
THE SPECTRAL PROPERTIES OF THE AGN TYPE 2 SAMPLE: THE SEARCH FOR THE SIGMA* SURROGATE 33

Daniela Kirilova and Mariana Panayotova
INFLATIONARY MODELS, REHEATING AND SCALAR FIELD CONDENSATE BARYOGENESIS 39

Jelena Petrovic
EVOLUTION OF MASSIVE BINARY SYSTEMS: PRIMARY STAR EVOLUTION INTO A NEUTRON STAR 43

Lyubov Marinkova, Todor V. Veltchev and Sava Donkov
ANALYSIS OF THE DENSITY DISTRIBUTION IN STAR-FORMING CLOUDS: EXTRACTION OF A SECOND POWER-LAW TAIL 51

Mariyana Bogdanova, Orlin Stanchev and Todor V. Veltchev
STUDY OF THE FRACTAL DIMENSIONS IN THE MOLECULAR CLOUD ROSETTE BY USE OF DENDROGRAM ANALYSIS 61

Orlin Stanchev, Todor V. Veltchev and Mariyana Bogdanova
TRACING THE LOCAL MORPHOLOGY OF THE MOLECULAR CLOUD ROSETTE USING MOLECULAR-LINE DATA 69

Ruslan Zlatev, Nikola Petrov, Tsvetan Tsvetkov, Emil Ivanov, Rositsa Miteva, Velimir Popov, Yoana Nakeva and Ljube Bojevski
SHADOW BANDS AND RELATED ATMOSPHERIC PHENOMENA REGISTERED DURING TOTAL SOLAR ECLIPSES 79

Miroslava Vukcevic and Luka Č. Popović
SOLITONS IN THE IONOSPHERE – ADVANTAGES AND PERSPECTIVES 85

Aleksandra Kolarski
STORM ACTIVITY OVER BALKAN REGION DURING MAY 2009 93

Aleksandra Nina, Milan Radovanović, Luka Č. Popović, Ana Černok, Bratislav P. Marinković, Vladimir A. Srećković, Anđelka Kovačević, Jelena Radović, Vladan Čelebonović, Ivana Milić Žitnik, Zoran Mijić, Nikola Veselinović, Aleksandra Kolarski and Alena Zdravković
ACTIVITIES OF SERBIAN SCIENTISTS IN EUROPLANET 107

Natalija Janc, Milivoj B. Gavrilov, Slobodan B. Marković, Vojislava Protić Benišek, Luka Č. Popović and Vladimir Benišek
MILUTIN MILANKOVIĆ AND ASSOCIATES IN THE CREATION OF THE "*KANON*" 123

Milan S. Dimitrijević and Aleksandra Bajić
MYTHOLOGICAL ORIGIN OF CONSTELLATIONS AND THEIR DESCRIPTION: ARATUS, PSEUDO-ERATOSTHENES, HYGINUS 129

Aleksandra Bajić and Milan S. Dimitrijević
A PAIR OF STEĆAKS FROM DONJA ZGOŠĆA 139

Александра Бајић
ВЕНЕРА У МИТОЛОГИЈИ ЈУЖНИХ СЛОВЕНА
VENUS IN THE MYTHOLOGY OF THE SOUTHERN SLAVES 155

Петар В. Вуца
СУНЧАНИ ВАЛЦЕР
SUNDIAL WALTZ 169

LIST OF PARTICIPANTS 181

AUTHORS' INDEX 185

PROGRAMME OF THE CONFERENCE 186

Proceedings of the XII Serbian-Bulgarian Astronomical Conference
(XII SBAC) Sokobanja, Serbia, September 25-29, 2020
Editors: L. Č. Popović, V. A. Srećković, M. S. Dimitrijević and A. Kovačević
Publ. Astron. Soc. "Rudjer Bošković" No 20, 2020, 5-14

GAIA DR3 AND SOME RESULTS OF SERBIAN-BULGARIAN COOPERATION

GORAN DAMLJANOVIĆ

Astronomical Observatory, Volgina 7, 11060 Belgrade, Serbia
E-mail: gdamljanovic@aob.rs

Abstract. After ESA Gaia DR1 (First Release, 14th September 2016) and Gaia DR2 (Second release, 25th April 2018) the Third Release is going to be available, soon. At the end of 2020, an early Gaia EDR3 is expected, and the Gaia DR3 catalog is expected after July 2021. Mainly, the Gaia EDR3 catalog will be consisting of improved astrometry and photometry, but the Gaia DR3 catalog contains the Gaia EDR3 and other data: mean radial velocities for stars (with atmospheric-parameter estimates), variable-star classifications with the epoch photometry, solar-system results (some orbital solutions, epoch observations, etc.), double and multiple stars, QSOs and results of extended objects, etc. Some results of the Serbian-Bulgarian cooperation in line with Gaia mission are presented.

1. INTRODUCTION

Gaia is well into its extended mission lifetime, now. The predecessor of Gaia, the Hipparcos (High Precision PARallax Collecting Satellite) of the European Space Agency - ESA, with ~118000 stars, changed the astronomy at the end of the last century (ESA 1997; van Leeuwen 2007), but the Gaia with ~1 billion stars and near 500000 extragalactic sources (Prusti 2012), from milliarcsec to microarcsec astrometry (plus the radial velocities Vr), is going to be a revolution in astronomy. It is a cornerstone mission of ESA, and the spacecraft was launched at the end of 2013. The start of normal operations was in mid-2014 (from August 2014, the data are useful for the Gaia Second release - DR2). The nominal Gaia operations phase was five-year for the observations, but now it is longer than the predicted period.

The plan was to observe the stars in our Milky Way galaxy: astrometry, photometry, and spectroscopy; plus, about 500000 extragalactic sources (Prusti 2012). The goal is to produce the Gaia Catalogue in optical domain, and to replace the International Celestial Reference Frame – ICRF. The ICRF is based on the VLBI observations in the radio domain. The accuracy of positions and parallaxes are at the level of accuracy far away from the ground ones; the Gaia is scanning the full sky with very high precision. The photometry for all astrometrically detected objects was done, and in spectroscopy the goal is to get the catalogue of Vr for ~150 million sources, etc. In stellar astrophysics, the improvement in the distances (using the improvement of parallaxes via Gaia observations) will allow to obtain models of stars at different steps of evolution. About the binary and multiple stars, the orbital parameters could be calculated with much higher precision using the Gaia high spatial resolution, and about the exoplanets, it is expected to find several thousands of systems. Also, the Gaia will be of importance in the case of faint rare objects as brown or white dwarfs. It is detecting not only point sources, but the solar system objects (with an improvement of asteroid ephemerides, to determine the masses of the objects, etc.), galaxies (the few million ones), quasars – QSOs (of importance for reference systems and fundamental physics), etc.

The period of the Gaia precession spin axis is 63 days because there are two Gaia fields of view (they are separated by a basic angle of 106.°5), and Gaia rotates around itself with a period of 6 hours. In line with that, for each observed object, it is collecting between 40 and 250 measurements during the five-year observations (Prusti 2012; Taris et al. 2018). The estimated Gaia end-of-life is still end-2024, but the process of extending the mission for the period 2023-2025 will start, soon. The data processing for the Gaia early release (EDR3) has almost finished, and the Third Gaia Release (DR3) or catalog is expected after July 2021. Both solutions are based on 34 months of Gaia observations, and feature the same source list.

2. FIRST AND SECOND GAIA DATA RELEASES

The First Release of the Gaia catalogue (DR1) was available since 14th Sept. 2016. It was the first step of the future Gaia celestial reference frame (Gaia CRF). The ICRF is based on the VLBI radio observations of extragalactic sources, and the Gaia CRF is going to link to the ICRF; the Gaia catalogue is the main goal of that ESA mission. The Gaia is collecting data for more than one billion of sources (the number of QSOs is about 500000); the G is the white-light photometry band of Gaia. It is determining the high accurate positions, proper motions and parallaxes (five-parameter astrometric solution). In the DR1, there are ~2 million stars using Tycho-Gaia solution, and it is based on the first observational period (~14 months). Also, in that solution, only data were published about flux time-series variability detection for Cepheids and RR Lyrae, but not for AGN and QSOs. More about the DR1 could be found in Taris et al. (2018).

The Second solution of the Gaia catalogue (DR2) was released on 25 April 2018.

The list with DR2 files is presented in Table 1. At the top of that list is the file with the main part of the DR2 catalog (five-parameter astrometric solution for ~1.5 billion stars). The main data of some star (positions RA_ICRS and DE_ICRS, parallax, proper motions pmRA and pmDE) are presented in Table 2.

The parallax values given in DR2 could be less than zero, which, though meaningless, is due to the calculation procedure (as some cases in the Hipparcos Catalog). The DR2 is based on ~21 months (1.75 yr or 640 days) with some interruptions (period 22nd August 2014 – 23rd May 2016) of Gaia operational phase, even the start of astronomical observations was in July 2014. The first month of Gaia observations was not included in the DR2 solution because the data quality is not high enough. That catalogue contains results for 1.693 billion sources in the G magnitude range 3 to 21. For 1.332 billion sources there are five astrometric parameters: positions, proper motions, and parallaxes. The reference epoch is J2015.5 = JD 2457206.375 TCB = 2 July 2015 at 21:00:00 TCB, and it is about half-way through the observational period used in the DR2 solution (that epoch is 0.5 yr later than the DR1 one). Because of its, there are some differences in the positional data between the DR1 and DR2 releases. The reference epoch was chosen to get minimal correlations between the positions and proper motions.

Table 1: The second Gaia data release (DR2).

VizieR

Simple Target | List Of Targets — Fast Xmatch with large catalogs or Simbad

Target Name (resolved by Sesame) or **Position**: Clear | J2000 | Target dimension: 2 arcmin | Radius / Box size

I/345	Gaia DR2 (Gaia Collaboration, 2018) Similar Catalogs 2018A&A...616A...1G ReadMe+ftp timeSerie acknowledge and cite Gaia DR2 1 annotation(s) (on 1 specific record(s))
1.I/345/gaia2	(c)Gaia data release 2 (Gaia DR2). (**Download** all Gaia Sources as VOTable, FITS or CSV here. **Query from the command line using find_gaia_dr2 available in cdsclient** here) (original column names in green) *(1692919135 rows)*
2.I/345/rvstdcat	(c)Mean radial velocities on absolute scale (original column names in green)**[timeSerie]** *(4813 rows)*
3.I/345/rvstdmes	Original ground-based radial velocity measurements (original column names in green) *(71225 rows)*
4.I/345/allwise	Allwise AGN Gaia DR2 cross-identification (aux_allwise_agn_gdr2_cross_id) (original column names in green) *(555934 rows)*
5.I/345/iers	IERS GaiaDR2 cross-identification (aux_iers_gdr2_cross_id) (original column names in green) *(2820 rows)*
6.I/345/cepheid	(c)Cepheid stars (vari_cepheid) (original column names in green)**[timeSerie]** *(9575 rows)*
7.I/345/rrlyrae	(c)RR Lyrae stars (vari_rrlyrae) (original column names in green) *(140784 rows)*
8.I/345/lpv	(c)Long Period Variable stars (vari_long_period_variable) (original column names in green) *(89617 rows)*
9.I/345/varres	(c)Variability classification results of all classifiers, identified by the classifierName column (vari_classifier_result) (original column names in green) *(363969 rows)*
10.I/345/shortts	Short-timescale sources (vari_short_timescale) (original column names in green) *(3018 rows)*
11.I/345/tsstat	Statistical parameters of time series, using only transits not rejected (vari_time_series_statistics) (original column names in green) *(550737 rows)*
12.I/345/numtrans	Calibrated FoV transit photometry from CU5, consolidated and provided by CU7 for variable stars in Gaia DR2 (epoch_photometry, part 1) (original column names in green)**[timeSerie]** *(550737 rows)*
13.I/345/transits	Calibrated FoV transit photometry for CU5, consolidated and provided by CU7 for variable stars in Gaia DR2 (epoch_photometry, part 2) (original column names in green) *(17712391 rows)*
14.I/345/rm	Rotation period in segment, part 1 (vari_rotation_modulation) (original column names in green) *(147535 rows)*
15.I/345/rmseg	Rotation period in segment, part 2 (vari_rotation_modulation) (original column names in green) *(583988 rows)*
16.I/345/rmout	Rotation period in segment, part 3 (vari_rotation_modulation) (original column names in green) *(990561 rows)*
17.I/345/ssoobj	*Data related to Solar System objects observed by Gaia (sso_source) (original column names in green) *(Note)* *(14099 rows)*
18.I/345/ssoorb	*Auxiliary information on asteroid orbits and basic photometric parameters (aux_sso_orbits) (original column names in green) *(Note)* *(14099 rows)*
19.I/345/ssores	*Residuals with respect to an orbital fit considering only the Gaia observations (aux_sso_orbit_residuals) (original column names in green) *(Note)* *(1977702 rows)*
20.I/345/ssoobs	(c)*Solar System object observations (sso_observation) (original column names in green) *(Note)* *(1977702 rows)*

The positions (RA_ICRS, DE_ICRS) and proper motions (pmRA, pmDE) refer to the ICRS; see Table 2. An overview of Gaia EDR3 is done in Table 3. There are the approximate RA_ICRS and DE_ICRS for an additional 0.361 billion mostly faint objects (Lindegren et al. 2018). The DR2 is available in the online Gaia Archive[1]. In DR2 (as in DR1 case) there are only photometric data as time-series for Cepheids and RR Lyrae, but no QSOs and other objects with unstable flux.

Table 2: The coordinates (RA_ICRS, DE_ICRS), parallax, proper motions (pmRA, pmDE), Gmag, and other results of Gaia DR2.

VizieR

Send to VO tools

- Show the target form
- Show constraint information

The 3 columns in *color* are computed by VizieR, and are *not part of the original data.*

I/345/gaia2 Gaia DR2 (Gaia Collaboration, 2018) 2018A&A...616A...1G ReadMe+ftp

1 annotation(s) · post Gaia data release 2 (Gaia DR2). (**Download** all Gaia Sources as VOTable, FITS or CSV here. **Query from the command line using find_gaia_dr2 available in cdsclient** here)

(original column names in green) *(1692919135 rows)*

start AladinLite plot the output query using TAP/SQL

Full	RA_ICRS deg	e_ mas	DE_ICRS deg	e_ mas	Source	Plx mas	e_ mas	pmRA mas/yr	e_	pmDE mas/yr	e_	Dup	FG e-/s	e_ e-/s	Gmag mag	e_ mag
☐	038.20256418497	0.2024	+20.28818994203	0.1823	86621076520103680	-0.0469	0.2274	-0.082	0.509	-0.616	0.432	0	1.4657e+03	1.7194e+01	17.7733	0.0127

Display Full Selected Rows

ng the VizieR Service
data

© Université de Strasbourg/CNRS
· Contact

The DR2 catalogue is independent because it does not include any other astrometric data (Hipparcos or Tycho ones), in contrast to the DR1 which is the Tycho-Gaia astrometric solution (Lindegren et al. 2016). All sources are reduced as single stars (as in DR1 solution) and the main values are presented by the five astrometric parameters, but the results of some binary stars refer to the photocentre (in the case of unresolved binaries) or to either component (for resolved ones). The models, algorithms, and the main steps of astrometric solution are described in Lindegren et al. (2012). Also, there are some additions since 2012. The color information in DR2 for most of the sources could be found; it was obtained by using the photometric processing of data via the blue BP and red RP photometers. For bright sources, it means G < 14 mag, the median uncertainty is near 0.04 mas in parallax and position (at J2015.5 – reference epoch of DR2). For G=17 mag, it is 0.1 mas, and 0.7 mas in the case of G=20 mag. About the pmRA and pmDE, the mentioned values are: 0.05 mas/yr, 0.2 mas/yr, and 1.2 mas/yr, respectively. The optical Gaia DR2 CRF is aligned with ICRS, and in line with Lindegren et al. (2018), it is non-rotating to within 0.15 mas/yr with respect to the QSOs. Using QSOs the primary solution was linked to the ICRS, and the astrometric calibration parameters of the CCDs were determined via an astrometric solution for 16 million selected objects, ~1% of the input data. Then, for the other sources astrometric parameters were calculated (Lindegren et al.

[1] https://archives.esac.esa.int/gaia

2018). Depending on positions, magnitude and color, the systematic effects in the parallaxes are less than 0.1 mas. The Barycentric Celestial Reference System – BCRS is the primary coordinate system: the origin is at the solar system barycentre, the axes are aligned with the ICRS, the barycentric coordinate time – TCB is the time-like coordinate of the BCRS. In line with Soffel et al. (2003), a consistent theory of relativistic astronomical reference systems was used during the processing of the Gaia data.

Table 3: The overview of Gaia EDR3[2].

	# sources in Gaia EDR3	# sources in Gaia DR2	# sources in Gaia DR1
Total number of sources	≈ 1,800,000,000	**1,692,919,135**	**1,142,679,769**
Number of 5-parameter sources	≈ 1,500,000,000	1,331,909,727	2,057,050
Number of 2-parameter sources	≈ 300,000,000	361,009,408	1,140,622,719
Sources with mean G magnitude	≈ 1,800,000,000	1,692,919,135	1,142,679,769
Sources with mean G_{BP}-band photometry	≈ 1,500,000,000	1,381,964,755	-
Sources with mean G_{RP}-band photometry	≈ 1,500,000,000	1,383,551,713	-
Gaia-CRF sources	≈ 1,500,000	556,869	2,191
Sources with radial velocities	expected with Gaia DR3 / ≈7,210,000 from Gaia DR2 in Gaia EDR3	7,224,631	-
Variable sources	expected with Gaia DR3 / see Gaia DR2	550,737	3,194
Known asteroids with epoch data	expected with Gaia DR3 / see Gaia DR2	14,099	-
Effective temperatures (T_{eff})	expected with Gaia DR3 / see Gaia DR2	161,497,595	-
Extinction (A_G) and reddening ($E(G_{BP}-G_{RP})$)	expected with Gaia DR3 / see Gaia DR2	87,733,672	-
Sources with radius and luminosity	expected with Gaia DR3 / see Gaia DR2	76,956,778	-
and more...	expected with Gaia DR3	-	-

The non-linear motions, the case of binary stars and other perturbations, are not included in DR2; only the uniform space motion of the object relative to the solar system barycentre was considered. The future Gaia release is going to include the mentioned motions (Lindegren et al. 2018). The DR2 is aligned with ICRS and non-rotating with respect to the QSOs as distant objects; the Gaia-CRF2 is the celestial reference frame of Gaia DR2. In a prototype version of ICRF3 there are 2843 sources (mostly QSOs), the optical counterparts of VLBI radio objects. The IAU Working Group "Third Realization of ICRF" is responsible for the ICRF3 catalogue which contains 4262 sources with accurate VLBI radio positions.

[2] https://www.cosmos.esa.int/web/gaia/earlydr3

3. THIRD GAIA DATA RELEASE (EDR3 AND DR3)

The Third Gaia Data Release is split into two releases: the early release (EDR3), and the full Gaia Data Release (DR3). The EDR3 will be consisting of: astrometric, photometric, and radial-velocity data, variable-star and non-single-star results, object classifications with multiple astrophysical parameters for stars, QSOs, galaxies, and unresolved binaries, exo-planets, epochs and transits for all objects; see Table 3. The early Gaia EDR3 is going to appear at the end of 2020. The Gaia DR3 catalog is expected after July 2021.

The DR3 solution (with improved astrometry and photometry of DR2 solution) will be consisting of: mean Vr velocities for stars without detected variability, object classification and astrophysical parameters (BP/RP and RVS spectra for spectroscopically objects), the epoch photometry and variable-star classifications, Solar-system results (preliminary orbital solutions, individual epoch observations), non-single star catalogues, etc.

Figure 1: The Belogradchik AO and famous rocks.

4. GAIA AND SERBIAN-BULGARIAN COOPERATION

A few years after the installation of the fist instrument (D=60 cm telescope, during 2011) at the Serbian new site (at the Astronomical Station Vidojevica – ASV) the Gaia satellite was launched (at the end of 2013) and during 2013 the local cooperation "Serbian-Bulgarian mini-network telescopes" was established; the Serbian-Bulgarian observational activities and investigation about the Gaia mission were started. The ASV belongs to Astronomical Observatory in Belgrade

– AOB[3]. At the two Bulgarian sites, Belogradchik (see Figure 1) and Rozhen, there are another 4 instruments of our interest. At Belogradchik it is the D=60 cm telescope, and at the Rozhen Observatory there are three instruments: the Schmidt – camera 50/70 cm, D=2 m, and D=60 cm telescopes. Since mid-2016 there is another ASV instrument (D=1.4 m telescope) via Belissima project[4] (see Figure 2). Using these 6 telescopes (Damljanović et al. 2014; Damljanović et al. 2018a; Damljanović et al. 2018b; Taris et al. 2018) we started astronomical observations in line with: the Gaia astrometry (QSOs useful for Gaia CRF and the link ICRF - Gaia CRF), Whole Earth Blazar Telescope - WEBT objects (mostly blazars), Gaia Alerts or Gaia-Follow-Up Network for Transients Objects (Gaia-FUN-TO), etc.

Our activities about the Gaia tasks are in line with the bilateral Serbian-Bulgarian joint research projects: "Observations of ICRF radio-sources visible in optical domain" during three-year period 2014-2016, "Study of ICRF radio-sources and fast variable astronomical objects" (2017-2019), and the actual one "Gaia Celestial Reference Frame (CRF) and fast variable astronomical objects" (2020-2022, the leader is G. Damljanović). These projects are within the framework between the Serbian Academy of Sciences and Arts – SASA (or SANU in Serbian language) and Bulgarian Academy of Sciences – BAS (or BAN in Bulgarian language).

The QSOs optical flux variations were investigated by using the original observations of QSOs (period 2013-2019), and it is in accordance with the future Gaia reference frame (Taris et al. 2018). That frame will be materialized by the optical positions of the objects, and it is going to link to the ICRF. The ICRF is based on the VLBI radio positions of mostly QSOs. Because of its, it is necessary to investigate flux variability of QSOs. The unstable flux of QSOs indicates changes in the source structure. It is of importance for the position of the target photocentre (also, the evolution in time of that center).

Flux variations of 47 objects, mostly QSOs (Bourda et al. 2008 and 2011) which are suitable for the mentioned link, are presented in (Taris et al. 2018) and are based on the data in line with the "Serbian-Bulgarian mini-network telescopes".

About the Gaia Alerts, during six-year period (Oct. 2014 – Oct. 2020) we observed ~90 objects (or ~3300 CCD images) using mentioned 6 telescopes; it is ~15 objects per year (or ~550 images per year). There are a few published papers about Gaia Alerts (Campbell et al. 2015; Damljanović et al. 2014), and some results were presented at a few conferences. We are doing with the Johnson BV and Cousins RcIc filters, and the seeing is between 1."0 and 3."5 at the Serbian-Bulgarian sites: the mean value at ASV is 1."2, and it could be 0."7 at Rozhen and ASV. Using the 2 m Rozhen and 1.4 m ASV telescopes we could observe the objects until V=20 mag with Exp.~5 min (until 19 mag using smaller instruments).

---

[3] http://vidojevica.aob.rs

[4] http://belissima.aob.rs

In line with an indication that the bright DR2 reference frame (stars with G≤13 mag) rotates by about 0.1 mas/yr relative to the faint quasar frame, the amplitude A=0.5±0.1 mas/yr of the sinusoidal curve of pmDE differences INDLS-DR2 was calculated (Damljanović 2020) using the original INDLS catalog of 682 stars.

5. CONCLUSIONS

The Gaia DR2 sources have full astrometric data (proper motions and parallaxes, also), and the DR2 contains a vastly increased number of sources in comparison with DR1. The data of bright sources (G < 12 mag) were included already in DR1, but the results in DR2 are generally more accurate, and the results of DR2 are independent of the Hipparcos and Tycho catalogues. The Gaia CRF2, reference frame, is defined by Gaia data of QSOs. Also, the optical counterparts of VLBI objects in a solution of the ICRF3 were used.

Figure 2: The 1.4 m ASV telescope in its dome.

The astrometric results in Gaia DR2 are based on ~2 years of observations, and the random and systematic errors are still higher than it can be expected for the final solution. The calibrations are very preliminary. There are possibilities for improvements (Lindegren et al. 2018), and there are many investigations to be done for the next solution (DR3). Also, all objects beyond the solar system are treated as point sources in DR2 and this is necessary to improve in the reduction model, there are the unresolved binaries (the photocentre was calculated instead of each separate star of the system), the systematic errors are mainly based on the analysis of QSOs data, etc. There are a lot of parts for improving Gaia DR3.

In line with Gaia the cooperation "Serbian-Bulgarian mini-network telescopes" was established during 2013, SANU-BAN project was started at 2014. Then, the investigations in accordance with Gaia astrometry (Taris et al. 2018) and Gaia Alerts (Campbell et al. 2015; Damljanović et al. 2014) were done. The first SANU-BAS project was "Observations of ICRF radio-sources visible in optical domain" (2014-2016), the second one was "Study of ICRF radio-sources and fast variable astronomical objects" (2017-2019), and the actual one is "Gaia Celestial Reference Frame (CRF) and fast variable astronomical objects" (2020-2022). During six-year period ~90 Gaia Alerts (or ~3300 CCD images) were observed.

The amplitude A=0.5± 0.1 mas/yr of the sinusoidal curve of pmDE differences INDLS-DR2 was obtained using the original INDLS catalog of 682 stars; it is in line with an indication that the bright reference frame of DR2 (stars with $G \leq 13$ mag) rotates by about 0.1 mas/yr relative to the faint quasar frame (Damljanović 2020). We continue activities and investigation in line with that ESA mission.

Acknowledgements

I acknowledge the observing grant support from the Institute of Astronomy and Rozhen NAO BAS through the bilateral joint research project "Gaia Celestial Reference Frame (CRF) and fast variable astronomical objects" (2020-2022, leader is G. Damljanović), and support by the Ministry of Education, Science and Technological Development of the Republic of Serbia (contract No 451-03-68/2020-14/200002).

References

Bourda G., Charlot P., Le Campion J.: 2008, *Astronomy and Astrophysics,* 490, 403.

Bourda G., Collioud A., Charlot P., et al.: 2011, *Astronomy and Astrophysics,* 526, A102.

Campbell H.C., et al.: 2015, *MNRAS,* 452, 1960.

Damljanović G., Vince O., Boeva S.: 2014, *Serb. Astron. J.,* 188, 85-93.

Damljanović G., Taris F., Andrei A.: 2018a, *Astronomy and Astrophysics in the Gaia sky Proc., IAU Sym. No.,* 330 (2017), A. Recio-Blanco, P. de Laverny, A.G.A. Brown and T. Prusti, eds., 88-89.

Damljanović G., Latev G., Boeva S., Vince O., Bachev R., Jovanović M.D., Cvetković Z., Pavlović R.: 2018b, *Publ. Astron. Obs. Belgrade,* 98, 277-280.

Damljanović G.: 2020, *Astron. Nachr.,* 341, 770-780.

ESA: 1997, *The Hipparcos and Tycho Catalogues, ESA SP-1200.*

Lindegren L., Lammers U., Hobbs D., et al.: 2012, *Astronomy and Astrophysics,* 538, A78.

Lindegren L., Lammers U., Bastian U., et al.: 2016, *Astronomy and Astrophysics,* 595, A4.

Lindegren L., Hernandez J., Bombrun A., et al.: 2018, *Astronomy and Astrophysics,* 616, A2.

Prusti T.: 2012, *Astron. Nachr.,* 333, No. 5/6, 453–459.

Soffel M., Klioner S.A., Petit G., et al. : 2003, *AJ,* 126, 2687.

Taris F., Damljanovic G., Andrei A., Souchay J., Klotz A., Vachier F. : 2018, *Astronomy and Astrophysics,* 611, A52.

van Leeuwen F.: 2007, *Hipparcos, the New Reduction of the Raw Data, Astrophysics and Space Science Library, ed..,* 350.

Proceedings of the XII Serbian-Bulgarian Astronomical Conference
(XII SBAC) Sokobanja, Serbia, September 25-29, 2020
Editors: L. Č. Popović, V. A. Srećković, M. S. Dimitrijević and A. Kovačević
Publ. Astron. Soc. "Rudjer Bošković" No 20, 2020, 15-22

SERBIAN-BULGARIAN OBSERVATIONS OF GAIA ALERTS (GAIA-FUN-TO) DURING 2019

GORAN DAMLJANOVIĆ[1], RUMEN BACHEV[2], SVETLANA BOEVA[2], GEORGY LATEV[2], MILAN STOJANOVIĆ[1], MILJANA D. JOVANOVIĆ[1], OLIVER VINCE[1], ZORICA CVETKOVIĆ[1], RADE PAVLOVIĆ[1] and GABRIJELA MARKOVIĆ[3]

[1]*Astronomical Observatory, Volgina 7, 11060 Belgrade, Serbia*
[2]*Institute of Astronomy with NAO, BAS, BG-1784, Sofia, Bulgaria*
[3]*Faculty of Mathematics, University of Belgrade, Studentski trg 16, 11000 Belgrade, Serbia*
E-mail: gdamljanovic@aob.rs, bachevr@astro.bas.bg, sboeva@astro.bas.bg, glatev@astro.bas.bg, mstojanovic@aob.rs, miljana@aob.rs, ovince@aob.rs, zorica@aob.rs, rpavlovic@aob.rs, gbesic@yahoo.com

Abstract. We used a set of six optical Serbian-Bulgarian telescopes at three sites (Belogradchik, Rozhen, and Vidojevica) to monitor astronomical objects in line with: the Gaia ESA mission (Gaia astrometry, Gaia Alerts or Gaia-Fellow-Up Network for Transients Objects), the Whole Earth Blazar Telescope - WEBT international project, cataclysmic and symbiotic stars, etc. Some results about observations of Gaia Alerts (Gaia-FUN-TO) during 2019 using "the Serbian-Bulgarian mini-network telescopes" are presented, here. Usually, we did about 15 objects per year. The mentioned activities are in line with actual SANU-BAN (Serbian and Bulgarian Academies of Sciences) joint research project "Gaia Celestial Reference Frame (CRF) and fast variable astronomical objects" (period 2020-2022), and similar international investigations supported by IAU. Also, our activities in line with the object Gaia18dvy are presented. The Gaia18dvy is a new case of FU Orionis-type young eruptive stars in the Cygnus OB3 Association, and the paper about it was published during 2020.

1. INTRODUCTION

The Gaia astronomical satellite of the European Space Agency – ESA is a space mission, and it is operating since mid-2014. Gaia is surveying the full sky: astrometrically, photometrically, and spectroscopically. It is doing revolution in astrometry, and these results are useful for all the relevant scientific communities:

for the stellar physics, our understanding of the Milky Way galaxy, the Solar system bodies, etc. The positions, proper motions and parallaxes (or the high-precision astrometric data) are the main goal of that mission. The G magnitudes of sources range from 3 to 21. An important step in the realization of the Gaia reference frame in future is the Gaia catalogue. The second Gaia solution or data release (DR2) of about 1.7 billion sources has been made publicly available on April 2018, and the next one (Gaia EDR3) is going to appear at the end of 2020.

The Gaia provides near-real-time photometric data during scanning the sky multiple times, and because of its these data are used to detect some changes in brightness from all over the sky (appearance of new objects, also). The next step is that the Gaia Science Alerts system produces alerts on the mentioned interesting objects, and we continue observations of these objects using the ground-based telescopes. The first alerts were published in October 2014 by the Gaia Photometric Science Alerts, and three years after that the Gaia Science Alerts was among the leading transient surveys in the world with transients as: supernovae, cataclysmic variables, microlensing events, other rare phenomena. More than 3000 transients were discovered until October 2017 (during three years).

2. GAIA ALERTS AND SERBIAN-BULGARIAN COOPERATION

Over the mission lifetime, Gaia has a goal of recording each object in the sky about 70 times. This produces a lot of alerts, where number of alerts increases with the number of observations of each object. Until the beginning of November 2020 about 14130 Gaia Alerts had been reported from all over the sky. In Table 1, we present a number of alerts published per year on the official website[1].

In 2011 the Serbian new astronomical site at the Astronomical Station of Vidojevica (ASV) of Astronomical Observatory in Belgrade (AOB) was established with a new D=60 cm telescope. At mid-2016 and via the Belissima project (see website[2]) there was another new telescope (D=1.4 m) at that site, and we used 4 instruments in Bulgaria (at Belogradchik and Rozhen sites) in line with our regional cooperation "Serbian-Bulgarian mini-network telescopes" which was started in 2013. Also, our activities are in line with the SANU-BAN joint research projects: "Observations of ICRF radio-sources visible in optical domain" (for the period 2014-2016), "Study of ICRF radio-sources and fast variable astronomical objects" (2017-2019), and the actual one "Gaia Celestial Reference Frame (CRF) and fast variable astronomical objects" (2020-2022, the leader is G.Damljanović). In a few papers (Damljanović et al. 2014; Taris et al. 2018) the main information about the mentioned instruments was published. Three telescopes (D=2 m, D=60 cm, and Schmidt-camera 50/70 cm) are at NAO Rozhen, and one (D=60 cm) is at Belogradchik AO. The NAO BAS means National Astronomical Observatory of

[1] http://gsaweb.ast.cam.ac.uk/alerts/home

[2] http://belissima.aob.rs

Bulgarian Academy of Sciences (or BAN in Bulgarian language), and Serbian Academy of Sciences and Arts is SASA (or SANU in Serbian language).

3. RESULTS

Our mini network of six telescopes (at three sites) is presented in Table 2, and the CCD cameras in line with these telescopes are presented in Table 3.

Also, we cooperate with colleagues from India, and because of this from time to time we can use D=1.31 m telescope of the Aryabhatta Research Institute of observational sciencES (ARIES). The ARIES is a site in the central Himalayan region (Manora Peak, Nainital); the geographic coordinates (longitude, latitude) and altitude are: λ=79.7° E, φ=29.4° N, h=2420 m. The mentioned telescope is equipped with the CCD camera Andor DZ436: 2048x2048 pixels, 13.5x13.5 μm pixel size, the scale is 0.″54 per pixel, FoV=18.′5x18.′5. This instrument is a modified R.-C. system Cassegrain; it is Devasthal Fast Optical telescope (DFOT).

Table 1: Number of Gaia alerts published per year.

Year	*Number of alerts*
2020 (updated 10th Nov 2020)	3468
2019	3915
2018	2729
2017	2322
2016	1522
2015	168
30th Aug 2014 – 31st Dec 2014	103

The standard bias, dark and flat-fielded corrections are done (also, hot/dead pixels are removed), and usually we did 3 CCD images per filter. The Johnson-Cousins BVRcIc filters were available. The Astrometry.Net and Source Extractor are used. The output is supposed to be submitted to the Cambridge Photometric Calibration Server (CPCS) for further calibration. About 3300 CCDs of the Gaia Alerts or Gaia-Follow-Up Network for Transients Objects (Gaia-FUN-TO) were collected during Oct. 2014 – Oct. 2020; it is about 550 images per year. About 90 objects were observed over the mentioned six years (it is near 15 objects per year).

Using our data there are a few published papers (Campbell et al. 2015; Wyrzykowski et al. 2020; Szegedi-Elek et al. 2020; Damljanovic et al. 2020; etc.).

In 2019, we observed 15 objects.

A) There are 9 objects using the **D=60 cm telescope at ASV**: Gaia19apc (5 times or epochs), Gaia19awc (2), Gaia19bcv (2), Gaia19cvu (1), Gaia19cup (1), Gaia19dke (1), Gaia19dum (2), Gaia18dvy (1), Gaia19duw (2).

B) Using the **D=1.4 m ASV** there are 10 objects: Gaia18dvu (2), Gaia18dvy (2), Gaia19aik (2), Gaia19ajp (2), Gaia19apc (3), Gaia19bcv (2), Gaia19awc (1), Gaia19drp (1), Gaia19dqe (1), Gaia19bpg (1).

C) Using the **D=2 m Rozhen** (FoReRo2 system) there are 3 objects: Gaia19awc (1), Gaia19bcv (1), Gaia19apc (1).
D) There are no data using the **D=60 cm Rozhen** (-): - .
E) There are no data using the **D=50/70 cm Schmidt-camera at Rozhen** (-): - .
F) There are no data using the **D=60 cm Belogradchik AO** (-): - .

We did about 440 CCDs during 2019: about 200 using D=60 cm ASV, 36 using D=2 m Rozhen (FoReRo2 system), and about 200 using D=1.4 m ASV. It is ~8% using D=2 m Rozhen, ~46% using 60 cm ASV, and ~46% using 1.4 m ASV.

Table 2: Mini network of telescopes.

Name	*Type*	*Longitude*	*Latitude*	*Altitude*
Astronomical Station Vidojevica – Astronomical Observatory Belgrade – (Serbia)				
"Milankovic" 1.4 m	Ritchey-Chrétien	21.6°	43.1°	1143 m
"Nedeljkovic" 60 cm	Cassegrain	21.5°	43.1°	1136 m
Rozhen National Astronomical Observatory – Bulgarian Academy of Sciences (Bulgaria)				
2 m	Ritchey-Chrétien	24.7°	41.7°	1730 m
60 cm	Cassegrain	24.7°	41.7°	1759 m
50/70 cm	Schmidt-camera	24.7°	41.7°	1759 m
Belogradchik Astronomical Observatory (Bulgaria)				
60 cm	Cassegrain	22.7°	43.6°	650 m

The object **Gaia18dvy** is very interesting. Its CCD image is presented (see Figure 1) after standard reduction (bias/dark/flat, hot/dead pixels, etc.). The object is marked with lines. This image was made on 13th March 2019 using the ASV telescope (D=1.4 m) with CCD Andor iKon-L camera: R-filter, Exp.=120s, FoV=8.′3x8.′3, binning=1x1, scale=0.″24 per pixel. We observed the Gaia18dvy three times during March and August 2019 using two ASV instruments:

1.) on 12th March, JD=2458555.6, using D=1.4 m ASV telescope with CCD Andor iKon-L (0.″244 per pixel), there are 12 CCD images or 3(BVRI),
2.) on 29th March, JD=2458572.6, D=1.4 m with Andor iKon-L, 3(BVRI),
3.) on 29th August, JD=2458725.5, D=60 cm with FLI PL230 (0.″518), 3(BVRI).

The obtained results are:

1.) in B-band it is 18.58 mag with st.dev.=0.03 mag (MJD=58555+0.0939 days), 18.55±0.03 (+0.1002), and 18.61±0.03 (+0.0876),
V=16.60±0.01 (+0.1018), 16.59±0.01 (+0.0955), 16.62±0.01 (+0.0892),
r=15.60±0.00 (+0.0971), 15.60±0.00 (+0.1034), 15.60±0.00 (+0.0907),
i=14.43±0.00 (+0.0986), 14.43±0.00 (+0.0923), 14.44±0.00 (+0.1049),

2.) in B-band it is 18.30±0.03 mag (58572+0.1158), 18.29±0.03 (+0.1219), and 18.32±0.03 (+0.1098),
V=16.38±0.01 (+0.1234), 16.38±0.01 (+0.1173), 16.37±0.01 (+0.1113),
r=15.42±0.00 (+0.1188), 15.42±0.00 (+0.1249), 15.40±0.00 (+0.1128),

i=14.34±0.00 (+0.1203), 14.34±0.00 (+0.1143), 14.33±0.00 (+0.1264),

3.) in B-band it is 17.54±0.04 mag (58724+0.9608), 17.57±0.04 (+0.9630), and 17.62±0.05 (+0.9587),

V=15.77±0.01 (+0.9694), 15.79±0.01 (+0.9673), 15.73±0.01 (+0.9651),
r=14.99±0.00 (+0.9737), 14.77±0.00 (+0.9758), 14.98±0.00 (+0.9715),
i=13.67±0.00 (+0.9801), 14.78±0.00 (+0.9779), 13.68±0.00 (+0.9822).

Table 3: CCD cameras of our mini network telescopes.

Telescope D/F [m]	*Camera*	*Chip size [pixel]*	*Pixel size [μm]*	*Scale ["]*	*Field of view - FoV[']*
1.4/11.42 ASV	Apogee Alta U42 Andor iKon-L	2048 x 2048 2048 x 2048	13.5 x 13.5 13.5 x 13.5	0.243 0.24	8.3 x 8.3 8.3 x 8.3
2/15.774 Rozhen	VersArray 1300B Andor iKon-L	1340 x 1300 2048 x 2048	20 x 20 13.5 x 13.5	0.261 0.176	5.6 x 5.6 6.0 x 6.0
0.6/6 ASV	Apogee Alta U42 SBIG ST10 XME	2048 x 2048 2184 x 1472	13.5 x 13.5 6.8 x 6.8	0.465 0.23	15.8 x 15.8 8.4 x 5.7
0.6/7.5 Rozhen	FLI PL09000	3056 x 3056	12 x 12	0.33	16.8 x 16.8
0.6/7.5 Belogradchik	FLI PL09000	3056 x 3056	12 x 12	0.33	16.8 x 16.8
0.5/0.7/1.72 Rozhen	FLI PL16803	4096 x 4096	9 x 9	1.08	73.7 x 73.7

Our results (suitable magnitudes) are in good accordance: with other presented results (see Figure 2), with the ground-based relative photometry, possibilities of our instruments, etc. These magnitudes are transferred from our set of filters (Johnson BV and Cousins RcIc) into another one via the Cambridge Server.

We investigated Gaia 18dvy (α=20:50:06.02, δ=36:29:13.52, see Figure 1.) which was noted by Gaia alerts system when its light curve had a 4 magnitude rise in the period 2018-2019. It was proved to be a new case of FU Orionis-type young eruptive stars in the Cygnus OB3 Association (Szegedi-Elek et al. 2020).

Once identified, Gaia18dvy was observed by Follow-Up-Network of telescopes in multi-national campaign. All follow-up data were collected and presented in the Figure 2. Here, we reproduce this as Figure 3 using the paper (Szegedi-Elek et al. 2020). Our images of Gaia 18dvy were among the first ones collected for this campaign and an example of our frame is shown in Figure 1, it is presented among other follow-up data in Figure 3. Our results fit very well other points of the light curve (Figure 3).

It was observed in optical and infrared domain which was followed by spectroscopic observations. Its optical and near-infrared spectroscopic characteristics in the outburst phase are consistent with those of FU Orionis-type young eruptive stars. The progenitor of the outburst is a low-mass K-type star. A radiative transfer modeling of the circumstellar structure has been developed, based on the spectral energy distribution, and indicates a disk with a mass of 4×10^{-3} M_{sun}. The known population of FU Orionis type stars is very small (only 26 FUors and FUor-like objects).

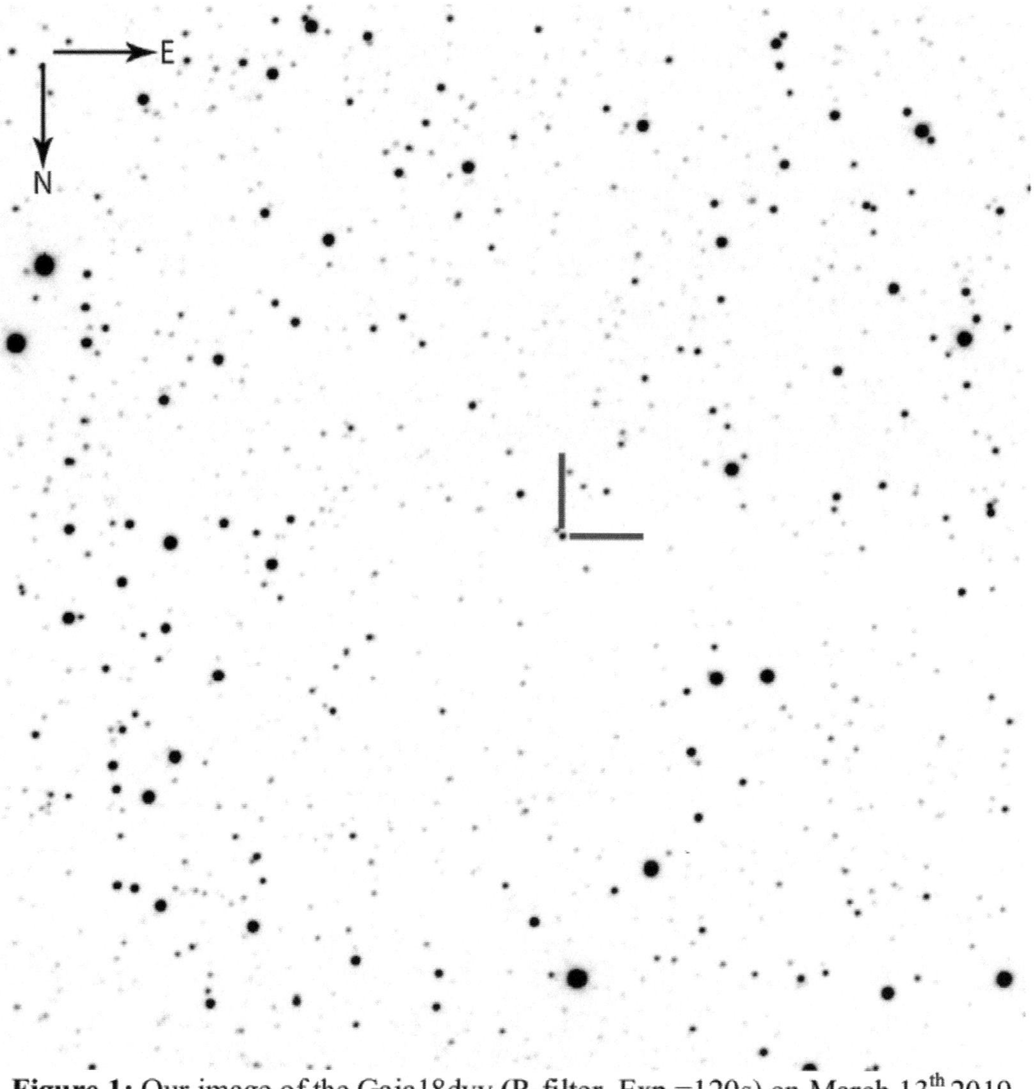

Figure 1: Our image of the Gaia18dvy (R-filter, Exp.=120s) on March 13th 2019 using the D=1.4 m ASV telescope with CCD Andor iKon-L camera.

It is possible to observe objects down to 20 mag in the V-band using D=2 m Rozhen and D=1.4 m ASV telescopes with Exp.=300s. With smaller telescopes, it is down to 19 mag. The D=1.4 m telescope is a new addition to our network from mid-2016 (new dome from 2018). In the last 2 years both bigger telescopes (2m Rozhen and 1.4 m ASV) have had new CCD cameras Andor iKon-L 936. The telescope D=2 m at Rozhen underwent realuminization of the mirrors in 2017.

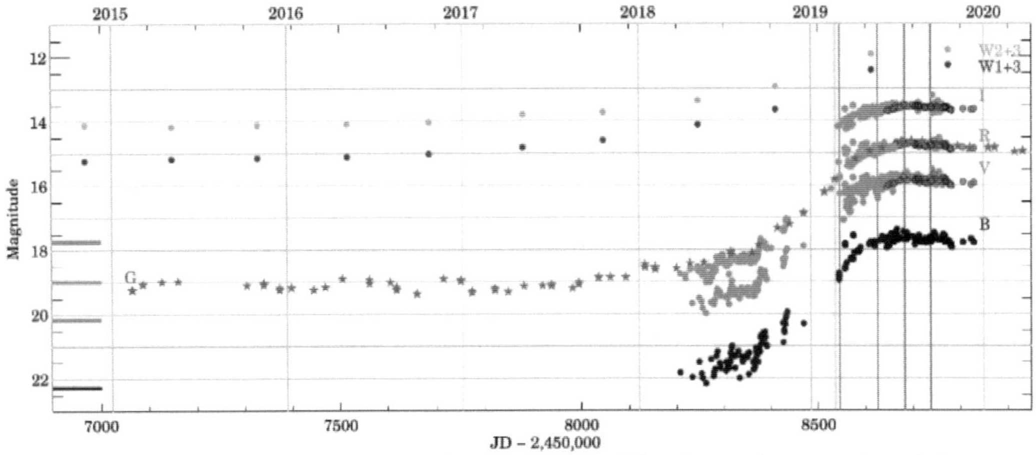

Figure 2: All follow-up data of Gaia 18dvy. The figure is reproduced from Szegedi-Elek et al. (2020) paper. The green asterisks show the Gaia data, the purple dots - WISE data, filled dots - ZTF and OPTICON data, the photometry from the Konkoly Observatory is highlighted by the black circles. Average Pan-STARRS magnitudes, converted to the Johnson-Cousins system, are indicated by the horizontal lines at the left side of the figure. The red vertical lines mark optical spectra of Gaia 18dvy, while the black vertical line indicates the epoch of NIR spectrum. The two blue vertical lines display the time period when the TESS satellite observed Gaia 18dvy.

4. CONCLUSIONS

The ESA Gaia astronomical satellite was launched in the end of 2013. In mid-2014 the first observations of that mission were done, and since October 2014 the Gaia Photometric Science Alerts started to publish alerts. During the first three years about 3000 alerts were issued by the Gaia Science Alerts group: cataclysmic variables, supernovae, candidate microlensing events, etc.

In line with our regional cooperation "Serbian-Bulgarian mini-network telescopes" and three SANU-BAN projects a few objects were observed (Damljanović et al. 2014) during the test phase in 2013 and 2014. After that (from the end of 2014), we continued the observations of the Gaia-Follow-Up Network for Transients Objects (Gaia-FUN-TO) or Gaia Alerts. From mid-2016 there were 6 Serbian-Bulgarian telescopes for the mentioned activities.

About 90 objects were observed during 6 years (until the end of October 2020), and it is near 15 objects per year. About 3300 CCD images were collected; it is about 550 images per year. Our observations were done in Johnson BV and Cousins RcIc filters. Usually, we did 3 CCD images per filter. At the beginning of Gaia Alerts the paper (Campbell et al. 2015) about rare object was published; it is the eclipsing AM CVn Gaia14aae object. During 2020, two papers were published: (Wyrzykowski et al. 2020) about Gaia16aye object (Ayers Rock), and (Szegedi-Elek et al. 2020) about Gaia18dvy one. The Gaia16aye is the binary microlensing event (the first discovered in the Northern Galactic Disk); we did that object from mid-2016 and during about 2.5 years. The object Gaia18dvy is a new case of FU Orionis-type young eruptive stars in the Cygnus OB3 Association. Also, some of our results were presented at a few conferences.

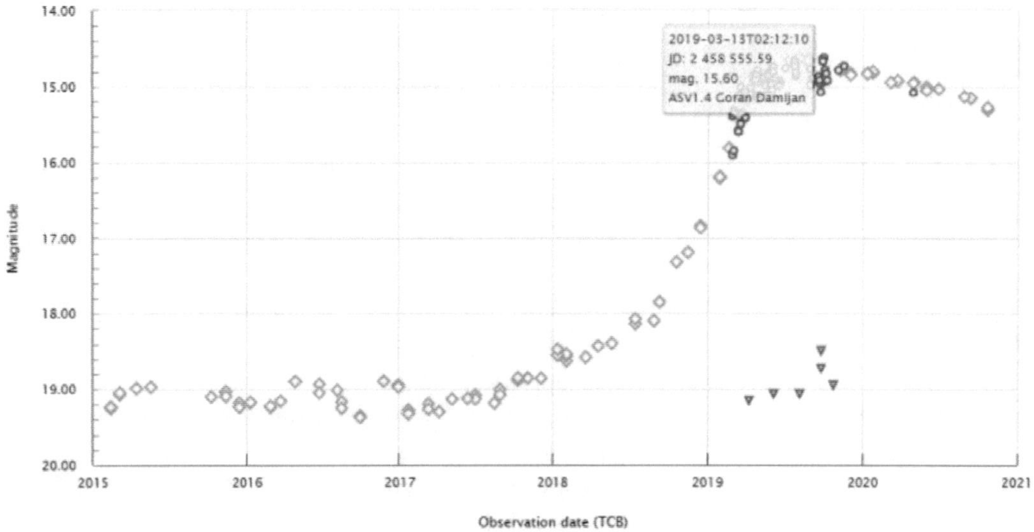

Figure 3. Our measurement of Gaia 18dvy (one shown as image in Figure 2) shown with the arrow, among other follow-up data in the campaign (brown circles) and Gaia data (blue diamonds).

Acknowledgements

We gratefully acknowledge the observing grant support from the Institute of Astronomy and Rozhen NAO (BAS), and support by the Ministry of Education, Science and Technological Development of the Republic of Serbia (contract No 451-03-68/2020-14/200002). We acknowledge ESA Gaia DPAC and Photometric Science Alerts Team (http://gsaweb.ast.cam.ac.uk/alerts).

References

Damljanović G., Vince O., Boeva S.: 2014, *Serb. Astron. J.*, 188, 85–93.
Damljanović G., et al.: 2020, *Bulgarian Astron. J.*, 32, 108–112.
Campbell H.C., et al.: 2015, *MNRAS*, 452, 1060-1067.
Szegedi-Elek E., et al.: 2020, *ApJ*, 899:130 (8pp).
Taris F., Damljanović G., Andrei A., et al.: 2018, *A&A*, 611, A52.
Wyrzykowski L., et al.: 2020, *A&A*, 633, A98.

Proceedings of the XII Serbian-Bulgarian Astronomical Conference
(XII SBAC) Sokobanja, Serbia, September 25-29, 2020
Editors: L. Č. Popović, V. A. Srećković, M. S. Dimitrijević and A. Kovačević
Publ. Astron. Soc. "Rudjer Bošković" No 20, 2020, 23-31

COLOR VARIABILITY OF SOME QUASARS IMPORTANT TO THE ICRF – GAIA CRF LINK

MILJANA D. JOVANOVIĆ, GORAN DAMLJANOVIĆ,
ZORICA CVETKOVIĆ, RADE PAVLOVIĆ and MILAN STOJANOVIĆ

Astronomical Observatory, Volgina 7, 11060 Belgrade, Serbia
E-mail: miljana@aob.rs, gdamljanovic@aob.rs,
zorica@aob.rs, rpavlovic@aob.rs, mstojanovic@aob.rs

Abstract. The International Celestial Reference Frame (ICRF) and Gaia CRF will be two reference frames with similar precision. To link ICRF provided with very long baseline interferometry (VLBI) in radio and Gaia CRF in optical wavelength is necessary to observe and monitor set of objects visible in both wavelengths. Observations of 47 candidate sources important for the link have been carried out by using two telescopes located at the Astronomical Station Vidojevica (of the Astronomical Observatory of Belgrade) and the one at the Rozhen National Astronomical Observatory (Bulgaria). The brightness variability (V and R bands) of five candidate sources was tested using the F-test. The color variability for the period from July 2016 to August 2019 for these five objects, and color magnitude dependences for the same period are presented in this paper.

1. INTRODUCTION

The third data realization of the International Celestial Reference Frame (ICRF3) was adopted by the International Astronomical Union (IAU) in August 2018 (Charlot et al. 2020). The ICRF3 is based on data obtained for about 40 years at radio frequencies (8.4 and 2.3 GHz) and data collected for the past 15 years at higher radio frequencies (24 GHz and dual-frequency 32 and 8.4 GHz) by very long baseline interferometry (VLBI).

The early third Gaia data release (Gaia EDR3) should be available from 3 December 2020, two years before the full data release. The Gaia EDR3 will contain data of about 1.8 billion sources and provide full astrometric information (positions, parallaxes, and proper motions) for about 1.5 billion sources. The currently available, second Gaia data release (Gaia DR2), published on 25 April 2018, provides celestial positions and G magnitudes for about 1.7 billion sources and provide positions, parallaxes, and proper motions for about 1.3 billion sources

(Gaia Collaboration et al. 2018). To link the Gaia CRF (based on the observations at optical wavelength) with the ICRF (based on the VLBI observations at radio wavelengths) it is required to observe a set of quasars (QSOs) visible in the optical domain. For this link are suitable only 10% of ICRF sources (about 70 sources). We observed the 47 candidates sources for this link with high astrometric quality (Bourda et al. 2011) since July 2016. These sources are Active Galactic Nuclei (AGNs), mostly QSO, the others are BL Lacertae – BL Lac and Seyfert galaxies type 1. The one of properties of AGNs is flux variability at different wavelength. These variations are divided into three classes (Gupta 2014): Intra Day Variabilities (those of less than a day), Short Term Variability (within the range of a few days to a few months), and Long Term Variability (from a few months to several years). It is necessary to monitor brightness variability of candidate sources for the link between two reference frames over a longer period of time.

The subject of this paper is investigation of color variability of five AGNs which have been the most observed objects. The results of investigation of the brightness variability of these five sources for a similar period of time are presented in papers: Jovanović (2019), and Jovanović and Damljanović (2020). The object properties are listed in Table 1: the coordinates are taken from SDSS DR14, the redshift (z) is from the NASA/IPAC Extragalactic Database – NED (https://ned.ipac.caltech.edu/), and the type is from SIMBAD Astronomical Database.

Table 1: The main properties of objects.

Object	$\alpha_{J2000.0}(^{o})$	$\delta_{J2000.0}(^{o})$	z	Type	Observation date (mmddyyyy) start	finish	No. of observations per filter V, R
1535+231	234.31041	23.01126	0.462524	QSO	07022016	08062019	20, 24
1556+335	239.72993	33.38851	1.653476	QSO	07022016	08062019	20, 27
1607+604	242.08560	60.30783	0.178000	BL Lac	07022016	08062019	23, 27
1722+119	261.26810	11.87096	0.018000	BL Lac	07022016	08082019	22, 26
1741+597	265.63334	59.75186	0.400000	BL Lac	07042016	08072019	34, 40

2. OBSERVATIONS AND PHOTOMETRY

The observations were made using three different telescopes. Two telescopes are located at Astronomical Station Vidojevica (ASV) of the Astronomical Observatory of Belgrade and third one is located at the Rozhen NAO in Bulgaria. The details about used telescopes (ASV with mirror diameter 60 cm and 1.4m, and Rozhen 2m) and mounted CCD cameras are presented in paper Jovanović (2019). Every nights two CCD images per V and R filter have been obtained and bias, dark and flat frames which were used for reduction of CCD images. The reduction and corrections (for bad pixels and cosmic rays) were done by using Image Reduction and Analysis Facility – the IRAF scripting language (ascl:9911.002) (Tody 1986, 1993).

The brightness of objects was determined using differential photometry with two comparison stars with similar brightness and color to that of the object. In the same manner, the brightness of a few control stars was determined for the every observing night. The (comparison and control) stars are located in the object vicinity. The stars were chosen from the Sloan Digital Sky Survey Data Release 14 (SDSS DR14) catalogue (Abolfathi et al. 2018), except for the object 1722+119 they are chosen from the set of stars proposed in paper Doroshenko et al. (2014). We selected stars following several criteria. Out of the stars in the objects vicinity, we chose stars which are not variable, not too bright or faint stars (with g, r and i magnitudes outside the range 14.5 – 19.5) or not very blue or red (outside the ranges $0.08 < r\text{-}i < 0.5$ and $0.2 < g\text{-}r < 1.4$), etc.

The transformation from the PSF g, r, and i magnitudes from the SDSS DR14 catalogue to the Johnson-Cousins V and R ones, was performed using equations (Chonis and Gaskel 2008):

$$V = g - (0.587 \pm 0.022)(g - r) - (0.011 \pm 0.013)$$
$$R = r - (0.272 \pm 0.092)(r - i) - (0.159 \pm 0.022).$$

The V and R magnitudes of the stars for the 1722+119 object are taken from the paper Doroshenko et al. (2014). In the Table 2. are listed coordinates of objects and their two comparison (A and B) and control stars, the V_C and R_C magnitudes of stars (obtained using mentioned equations for stars from SDSS DR14, and those from Doroshenko et al. (2014)), along with V_O and R_O (calculated magnitudes using differential photometry) of the objects and stars.

The presented objects were observed for about three years. During this period, brightness of two objects, 1722+119 and 1741+597, changed about 2.0 and 1.7 magnitudes in both filters, respectively. The brightness of these two objects has the higher standard deviation. The object 1722+119 has standard deviation around 0.6 mag and 1741+597 around 0.4 mag per filter. The light curve of R magnitude of object 1722+119 with light curves of its comparison and control stars is presented in Figure 1. The brightness of 1535+231 and 1607+604 had changes about 0.6 and 0.4 magnitudes in both filters, respectively. The standard deviations are around 0.2 and 0.1, per filter, respectively. The one object 1556+335 has the most stable brightness, the standard deviations are of the order of about of 0.01, similar to the standard deviations of stars. The calculated values of V_C and R_C (input values for differential photometry) magnitudes of stars are in good agreement with V_O and $R_{O,}$ and the standard deviations of brightness of comparison stars are similar to the ones of control.

In the paper Jovanović (2019) are presented charts of the fields of the objects and their (comparison and control) stars. In paper Jovanović and Damljanović (2020) are presented analysis of objects brightness variability and calculated amplitudes of their quasyperiods form the similar data sets.

Table 2: The coordinates, V and R magnitudes with standard errors of objects and their comparison and control stars; period July 2016 — August 2019.

Object No.	$\alpha_{J2000.0}(^o)$	$\delta_{J2000.0}(^o)$	$V_C \pm \sigma_{V_C}$(mag)	$R_C \pm \sigma_{R_C}$(mag)	$V_O \pm \sigma_{V_O}$(mag)	$R_O \pm \sigma_{R_O}$(mag)
1535+231	234.31041	23.01126			18.375 ±0.197	18.099 ±0.214
2 (A)	234.31491	23.01831	17.200 ± 0.031	16.658 ± 0.038	17.213 ±0.024	16.693 ±0.036
3	234.30004	23.02486	15.983 ± 0.030	15.633 ± 0.031	16.000 ±0.024	15.656 ±0.030
4 (B)	234.25178	23.01917	16.232 ± 0.024	15.867 ± 0.029	16.227 ±0.010	15.851 ±0.017
7	234.29312	22.96096	16.470 ± 0.027	15.973 ± 0.036	16.451 ±0.026	15.958 ±0.021
8	234.35917	23.01592	15.860 ± 0.035	15.149 ± 0.050	15.841 ±0.024	15.142 ±0.028
1556+335	239.72993	33.38851			17.501 ±0.048	17.020 ±0.040
2 (A)	239.71950	33.39110	17.336 ± 0.030	16.850 ± 0.038	17.344 ±0.031	16.895 ±0.032
3 (B)	239.69035	33.40959	16.381 ± 0.027	16.095 ± 0.030	16.378 ±0.013	16.074 ±0.015
5	239.76798	33.38778	16.271 ± 0.030	15.916 ± 0.031	16.289 ±0.025	15.936 ±0.023
6	239.74562	33.39003	16.198 ± 0.030	15.825 ± 0.031	16.225 ±0.022	15.876 ±0.021
7	239.74317	33.37370	15.552 ± 0.030	15.188 ± 0.031	15.568 ±0.022	15.223 ±0.017
8	239.73398	33.37219	15.743 ± 0.040	14.897 ± 0.064	15.756 ±0.045	14.966 ±0.016
1607+604	242.08560	60.30783			17.461 ±0.104	17.000 ±0.084
2 (A)	242.02882	60.28951	17.068 ± 0.027	16.619 ± 0.031	17.057 ±0.015	16.601 ±0.017
3 (B)	242.02526	60.31162	16.864 ± 0.025	16.423 ± 0.032	16.874 ±0.013	16.439 ±0.015
4	241.97352	60.35552	15.195 ± 0.025	14.781 ± 0.031	15.164 ±0.059	14.715 ±0.050
5	242.09638	60.34816	15.630 ± 0.031	14.965 ± 0.044	15.601 ±0.041	14.929 ±0.036
7	242.16854	60.37746	16.856 ± 0.024	16.467 ± 0.031	16.838 ±0.050	16.389 ±0.061
1722+119	261.26810	11.87096			15.540 ±0.633	15.013 ±0.613
C2	261.27167	11.86997	13.173 ± 0.005	12.570 ± 0.006	13.177 ±0.019	12.628 ±0.025
C3	261.24375	11.86636	14.078 ± 0.012	13.600 ± 0.008	14.082 ±0.020	13.629 ±0.015
1	261.31208	11.89125	13.445 ± 0.009	12.848 ± 0.010	13.440 ±0.024	12.857 ±0.024
2 (A)	261.30458	11.86519	14.823 ± 0.008	14.691 ± 0.012	14.829 ±0.008	14.688 ±0.005
5	261.25667	11.91311	15.873 ± 0.010	15.385 ± 0.016	15.857 ±0.043	15.378 ±0.022
9	261.23333	11.87083	15.809 ± 0.008	15.332 ± 0.014	15.799 ±0.019	15.339 ±0.018
10	261.23875	11.87083	16.142 ± 0.011	15.699 ± 0.019	16.134 ±0.013	15.712 ±0.020
C4 (B)	261.28958	11.85344	15.665 ± 0.009	15.164 ± 0.013	15.653 ±0.017	15.169 ±0.007
1741+597	265.63334	59.75186			17.945 ±0.372	17.545 ±0.376
2	265.62329	59.75176	15.565 ± 0.029	15.204 ± 0.054	15.619 ±0.032	15.290 ±0.042
3 (A)	265.57081	59.75387	16.673 ± 0.029	16.314 ± 0.053	16.678 ±0.014	16.331 ±0.015
4	265.68412	59.76861	16.376 ± 0.034	15.795 ± 0.067	16.407 ±0.029	15.840 ±0.024
5	265.61457	59.79547	16.154 ± 0.031	15.704 ± 0.056	16.187 ±0.040	15.753 ±0.024
6	265.68288	59.71901	16.126 ± 0.038	15.684 ± 0.064	16.137 ±0.023	15.696 ±0.022
7 (B)	265.59766	59.71686	16.633 ± 0.039	16.124 ± 0.074	16.629 ±0.013	16.110 ±0.012

Note. (A), (B) refer to comparison stars.

3. METHODS AND RESULTS

Some of the data were rejected after applying 3-σ rule. After implementation of Shapiro-Wilk test of normality we conclude that statistical tests which require normal data distribution can be applied.

3.1 F-test

We used the F-test to investigate the brightness variability of objects (de Diego 2010, Gupta et al. 2017, Jovanović 2019), by comparing variances of two data sets. The test statistic are:

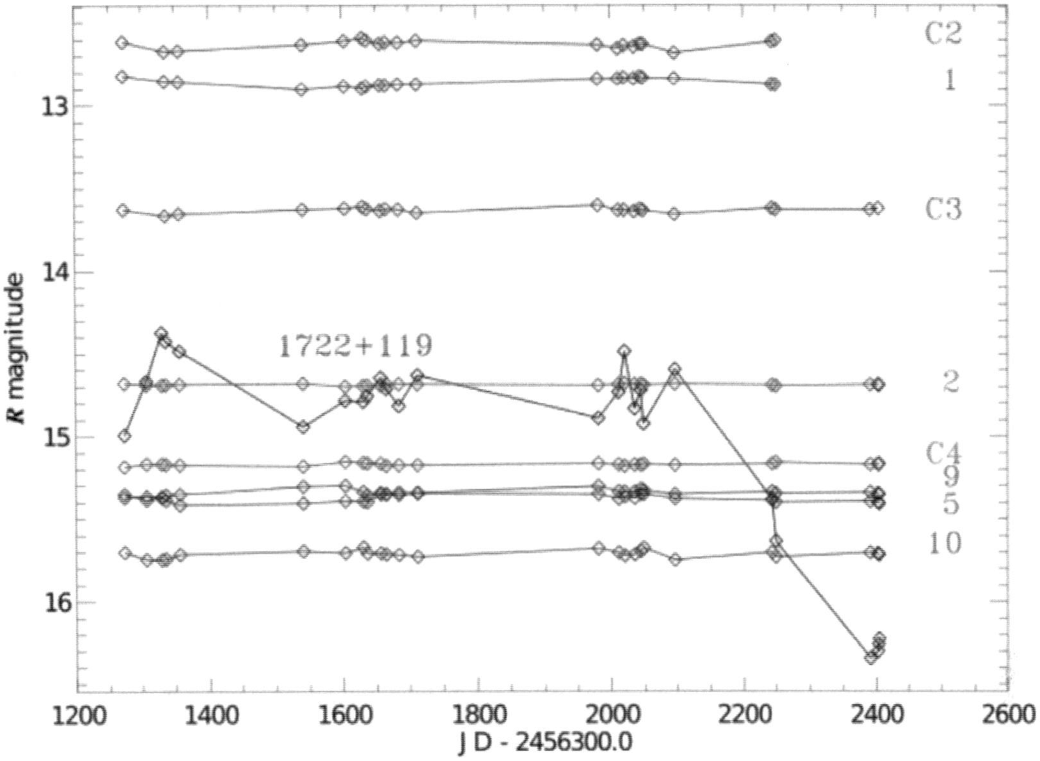

Figure 1: The light curve of R magnitude of 1722+119 and light curves of its comparison and control stars, from July 2016 until August 2019.

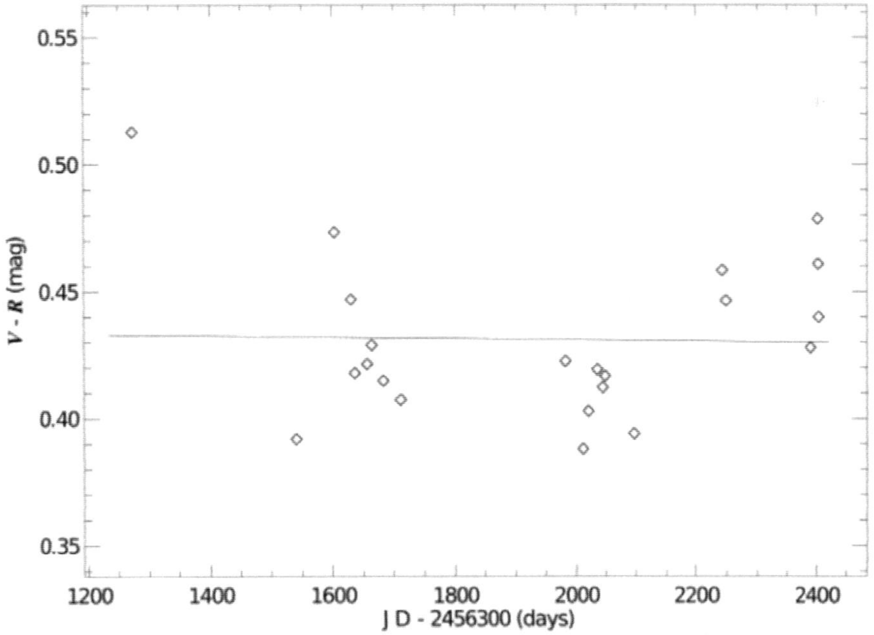

Figure 2: The light curve of color indices V-R variability during period July 2016 – August 2019, of object 1722+119.

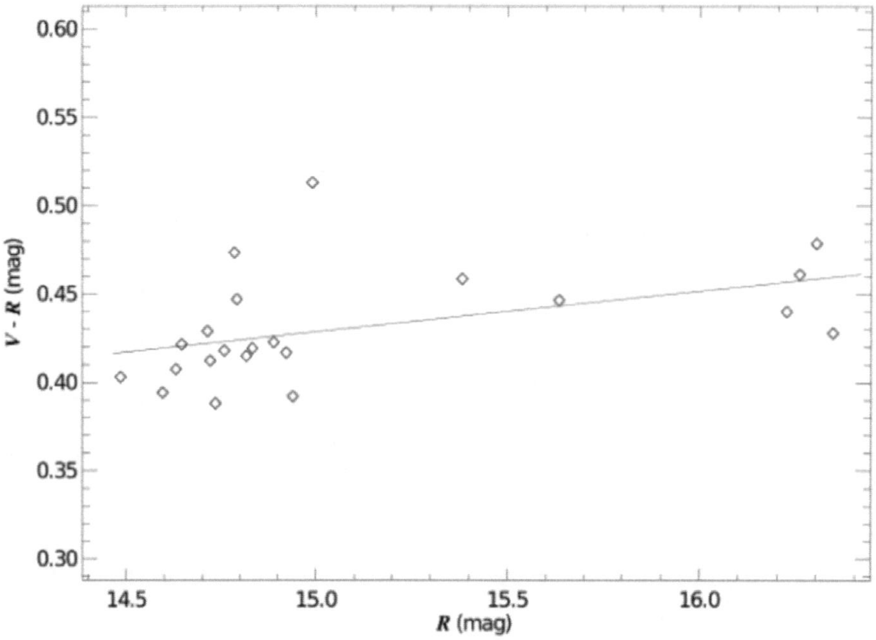

Figure 3: The correlation between color indices V-R and R-band magnitude, of object 1722+119.

$$F_1 = \frac{VAR(O-A)}{VAR(O-B)}, \quad F_1 = \frac{VAR(O-A)}{VAR(A-B)} \quad \text{and} \quad F_1 = \frac{VAR(O-B)}{VAR(A-B)}.$$

The designations O − A, O − B, and A − B refer to the differences of obtained magnitudes of object (O) and comparison star A, object and comparison star B, and comparison star A and B, respectively. The brightness of control stars was investigate by using the same test statistics (O refers to the magnitude of the control star). The variances VAR(O-A), VAR(O-B) and VAR(A-B) are the variances of mentioned differences.

The F_i (i=1, 2, 3) statistics were compared with critical values. The F_1 value should be around 1, because it is expected that the variances VAR(O-A) and VAR(O-B) should be close to each other. Tested brightness should be variable in the same manner for both comparison stars (A and B). When the F_2 and F_3 values are greater then critical (which correspond to the significance level 0.05 and number of freedom N-1, where N is the number of data), the null hypothesis (non variability) is discarded.

Results of F-test are different from in paper Jovanović and Damljanović (2020) because of the changing the selection of comparison stars (1607+604), and after rejecting some CCD images which were obtained during bad atmospheric conditions (1722+119 and 1741+597). The test shows that one object 1556+335 is non variable, the F_i values are almost equal 1, for V (F_1=1.11, F_2=1.34, and F_3=1.49) and R filter (1.23, 1.22, and 1.01). The critical values are 2.17 and 1.93, for V and R filter respectively. For the four variable objects F_1 values are around 1 as it is expected, but F_2 and F_3 are greater than the critical value. The $F_{2,3}$ values, along with number of data points N, and critical values Fc (for N and $\alpha = 0.05$) are listed in Table 3, for both filters. It is noticeable that the objects with significant brightness variability (1722+119 and 1741+597) have the higher $F_{2,3}$ values. The test shows that the brightness of control stars could be considered as non-variable.

Table 3: The F-test results.

Object	Filter	N	F_2, F_3	F_c
1535+231	V	20	31.39, 36.23	2.17
	R	24	15.08, 17.14	2.01
1607+604	V	23	14.09, 14.10	2.05
	R	27	6.69, 7.02	1.93
1722+119	V	22	621.18, 596.30	2.08
	R	26	2833.73, 2852.28	1.96
1741+597	V	34	190.86, 196.33	1.79
	R	40	183.66, 187.02	1.70

3.2 Color variability

We investigate the color indices V-R variations with respect to time (color vs. time) and with respect to R-band magnitude (color versus magnitude) of the five presented objects (variable and non variable). The data are fitted using Linear Least Squares (LLS) fit (y=a+bx). In addition, the linear Pearson correlation coefficients, r, were calculated

$$r = \frac{\sum(x-\bar{x})(y-\bar{y})}{\sqrt{\sum(x-\bar{x})^2 \sum(y-\bar{y})^2}}$$

where x is the time (color – time dependences) or R magnitude (color – magnitude dependences), y is color, x and y are corresponding average values of data.

The values of the standard errors of the (LLS) fits, intercepts, a, the slopes, b, the linear Pearson correlation coefficients, r, for color vs. time and color vs. magnitude are reported in Tables 4 and 5, respectively.

Table 4: The color variations with respect to time.

Object	σ_0	$a \pm \sigma_a$	$b \pm \sigma_b$	r
1535+231	0.084	0.347 ± 0.111	-5.36E-5 ± 5.88E-5	-0.205
1556+335	0.052	0.496 ± 0.072	-6.63E-6 ± 3.85E-5	-0.039
1607+604	0.067	0.587 ± 0.092	-6.02E-5 ± 4.82E-5	-0.258
1722+119	0.032	0.436 ± 0.041	-2.56E-6 ± 2.06E-5	-0.027
1741+597	0.085	0.649 ± 0.107	-8.41E-5 ± 5.53E-5	-0.260

Note. σ_0 =*stanadrd error*, a = *intercept*, b = *slope* of Color indices against $JD - 2456300$ and r = *Pearson coefficient.*

Table 5: The color – magnitude dependences.

Object	σ_0	$a \pm \sigma_a$	$b \pm \sigma_b$	r
1535+231	0.076	3.468 ± 1.380	-0.178 ± 0.076	-0.472
1556+335	0.047	10.300 ± 4.441	-0.577 ± 0.261	-0.452
1607+604	0.069	-0.788 ± 3.307	0.074 ± 0.195	0.081
1722+119	0.028	0.084 ± 0.148	0.023 ± 0.010	0.456
1741+597	0.085	-0.552 ± 0.743	0.060 ± 0.043	0.240

Note. σ_0 =*stanadrd error*, a = *intercept*, b = *slope* and r = *Pearson coefficient.*

It seems that there are no consistent systematic trends in the color variations, as shown in Table 4 and LLS fits to the color versus time plots (one example is presented in Figure 2). The three BL Lac objects (1607+604, 1722+119, and 1741+597) show bluer when brighter (BWB) trends (Table 5), unlike two objects which are QSO type (1535+231 and 1556+335).

The examples of the color versus time and color versus magnitude plots, of the object 1722+119, are presented in Figure 2 and Figure 3, respectively.

4. CONCLUSIONS

In this paper are presented results of the color variation of five objects (3 BL Lac and 2 QSO) during a period of about 3 years. The calculated V_C and R_C magnitudes (used values for the differential photometry) of all comparison and control stars are in a good agreement with magnitudes which were determined from the observations (V_O and R_O) in line with their standard errors (see Table 2). We applied F–test to investigate brightness variability of objects and their control stars. The test shows that significant changes of brightness variability of stars were not detected, and we consider that they are suitable for photometric measurements. During period July 2016 – August 2019, only one object, 1556+335, did not show brightness variability. In the previously published papers (Jovanović 2019, Jovanović and Damljanović 2020) are presented amplitude of the quasyperiods of four variable objects. In this paper in the Table 3 are presented results of F-test for four variable objects. The two objects with higher differences between maximum and minimum brightness have the higher test statistics than the other objects. For the five presented objects in Table 4 and Table 5 are listed results of analysis of brightness color variations and color – R magnitude dependences. Two objects of QSO type were found redder when they are brighter trends and for three BL Lac objects bluer when they are brighter (BWB) trends. The BWB trend is a well-observed feature in blazars especially in BL Lac object. To better determine coefficients of fitting and understand physical process and cause of objects variability, we need to continue with observations. It is necessary to proceed with further observations in order to investigate variability of object brightness (the Intra Day and Long Term variability) and color.

Acknowledgements

During the work on this paper authors were financially supported by the Ministry of Education, Science and Technological Development of the Republic of Serbia through the contract No 451-03-68/2020-14/200002. We gratefully acknowledge the observing grant support from the Institute of Astronomy and Rozhen National Astronomical Observatory, Bulgarian Academy of Sciences, via bilateral joint research projects "Gaia Celestial Reference Frame (CRF) and fast variable astronomical objects", and "Astrometry and photometry of visual double and multiple stars".

References

Abolfathi B., Aguado D. S., Aguilar G., et al.: 2018, *Astrophys. J. Suppl. Ser.*, 235, 42.

Bourda G., Collioud A., Charlot P., Porcas R., Garrington S.: 2011, *Astron. Astrophys.*, 526, A102.

Charlot P., Jacobs C. S., Gordon D., et al.: 2020, arXiv e-prints arXiv:2010.13625, 2010.13625.

Chonis T. S., Gaskell C. M.: 2008, *The Astronomical Journal*, 135, 264.

De Diego J. A.: 2010, *Astron. J.*, 139, 1269.

Doroshenko V. T., Efimov Yu. S., Borman G. A., Pulatova N. G.: 2014, *Astrophysics*, 57, 2.

Gaia Collaboration, Brown A. G. A, Vallenari A., Prusti T., et al.: 2018, *A&A*, 616, A1.

Gupta A. C.: 2014, *J. Astrophys. Astr*., 35, 307.

Gupta A. C., Agarwal A., Mishra A., et al.: 2017, *Mon. Not. R. Astron. Soc.*, 465, 4423.

Jovanović M. D.: 2019, *Serb. Astron. J.*, 199, 55-64.

Jovanović M. D., Damljanović G.: 2020, *Bulg. Astron. J.*, Vol 33, 9pp.

Tody, D.: 1986, *Proceedings SPIE Instrumentation in Astronomy VI,* ed. D.L. Crawford, 627, 733.

Tody, D.: 1993, *Astronomical Data Analysis Software and Systems II, A.S.P. Conference Ser*., eds. R.J. Hanisch, R.J.V. Brissenden, & J. Barnes, 52, 173.

Proceedings of the XII Serbian-Bulgarian Astronomical Conference
(XII SBAC) Sokobanja, Serbia, September 25-29, 2020
Editors: L. Č. Popović, V. A. Srećković, M. S. Dimitrijević and A. Kovačević
Publ. Astron. Soc. "Rudjer Bošković" No 20, 2020, 33-38

THE SPECTRAL PROPERTIES OF THE AGN TYPE 2 SAMPLE: THE SEARCH FOR THE SIGMA* SURROGATE

JELENA KOVAČEVIĆ-DOJČINOVIĆ[1], IVAN DOJČINOVIĆ[2],
MAŠA LAKIĆEVIĆ[1] and LUKA Č. POPOVIĆ[1]

[1]*Astronomical Observatory, Volgina 7, 11060 Belgrade, Serbia*
[2]*Faculty of Physics, University of Belgrade, Studentski trg 12, 11000, Belgrade, Serbia*
E-mail: jkovacevic@aob.rs

Abstract. M_{BH} - σ^* relation is the most commonly used method for the M_{BH} (mass of the super-massive black hole) estimation in the AGNs (Active Galactic Nuclei) Type 2. Since σ^* (stellar velocity dispersion) is in some cases diluted by the noise in AGN Type 2 spectra, it is of interest to find appropriate surrogate for σ^* in some prominent AGN Type 2 spectral characteristics. Here we used the large sample of the AGN Type 2 spectra from SDSS and try to find under what circumstances the width of the [O III]5007 Å emission lines can be used as σ^* surrogate. We find that only in the case of objects with no asymmetry in the narrow emission lines profiles, the width of the [O III] core can be used appropriately as σ^* surrogate, since outflow kinematics do not affect significantly the [O III] profile.

1. INTRODUCTION

According the Standard Unified model (Antonucci 1993, Urry & Padovani 1995S), in the center of an Active Galactic Nuclei there is a super-massive black hole (M_{BH}>10^6 Msun) surrounded by the accretion disc and high velocity gas of the Broad Line Region (BLR), where broad emission lines arise. Around BLR there is a torus of the dust, and out of the dusty torus, there is the Narrow Line Region (NLR), where the narrow emission lines arise. The AGNs Type 2 are observed through dusty torus (edge-on), which covers the broad emission lines from BLR, so only the narrow emission lines can be seen in their spectra. On the other hand, in the spectra of the AGNs Type 1, which are observed under higher inclination angle, both the narrow and the broad emission lines can be observed.

The estimation of the M_{BH} is of the great importance for investigation of the AGNs, their structure and evolution, as well as co-evolution with the host galaxy.

The commonly used method for M_{BH} estimation is M_{BH} - σ* relation (see e.g. Tremaine et al. 2002), where σ* is the velocity dispersion of the stars located in the bulge of the host galaxy, which are strongly gravitationally bounded to the M_{BH}. σ* is measured using the width of the stellar absorption lines in AGN spectra. The problem is that σ* is not always well measurable in Type 1 AGNs spectra due to strong luminosity of the AGN continuum, while in Type 2, it could be diluted by the noise. Taking into account that M_{BH} - σ* relation is the only method for the estimation of the M_{BH} for the Type 2 AGNs, it would be very useful to find some prominent spectral property that can be used as the surrogate for the σ*.

There were numerous attempts to find the surrogate for the σ* in Type 1 AGNs, using the width of the narrow emission lines (see Nelson 2000, Shields et al. 2003, Grupe & Mathur 2004, Greene & Ho 2005, Bonning et al. 2005, Salviander & Shields 2013). However, setback for larger application of this method is the large scatter between σ* and width of the narrow emission lines of Type 1 AGNs.

Nelson & Whittle (1996) noticed that there is lower correlation between σ* and the [O III] 5007 Å wings, than with the [O III] 5007 core. Recently, several analysis of the complex NLR kinematical structure are done, using the large samples of the AGNs (Woo et al. 2016, Eun et al. 2017). They showed that the narrow emission lines can be decomposed to the two components: one is the core component of the line, which arises from gravitationally bounded gas, and the other is the non-gravitational, wing component of the line, which arises in the gas outflow. It is of the great importance to correctly separate the gravitational from non-gravitational narrow line component, in order to find the good surrogate for the σ*.

In this work we analyze the sample of the AGNs Type 2, in order to find the subsample in which the width of the narrow lines can be used as appropriate surrogate for the σ*.

2. THE SAMPLE AND ANALYSIS

For this research, we obtained the sample of the 525 AGNs Type 2, taken from the SDSS, DR14, with high signal-to-noise ratio, S/N>20. After reddening and redshift correction, we performed spectral principal component analysis in order to subtract host galaxy contribution (see procedure in Lakićević et al. 2017).

We fit the [O III]λλ 4959, 5007 Å narrow lines with two component model: one Gaussian for the core, and another for the wing (see Fig 1). We found that in 26% of objects from the sample there is no any asymmetry in the [O III] line profiles, i.e. [O III] emission lines can be fitted with only one, core Gaussian component (see Fig 1, left). In the rest of the sample (~74%), the asymmetry is present in the [O III] profiles, so these emission lines are fitted with two Gaussians (core + wing component, see Fig 1, right).

Therefore, the initial sample of the 525 AGNs Type 2 is divided into two subsamples according to the strength of the outflow which is reflected in the strength (width and shift) of the wing components of the narrow emission lines. The first subsample consists of the 135 Type 2 AGN spectra, in which there are no wing components in narrow emission lines, and the other subsample consists of 391 AGN spectra, in which the wing (outflow) components are present in [O III] lines.

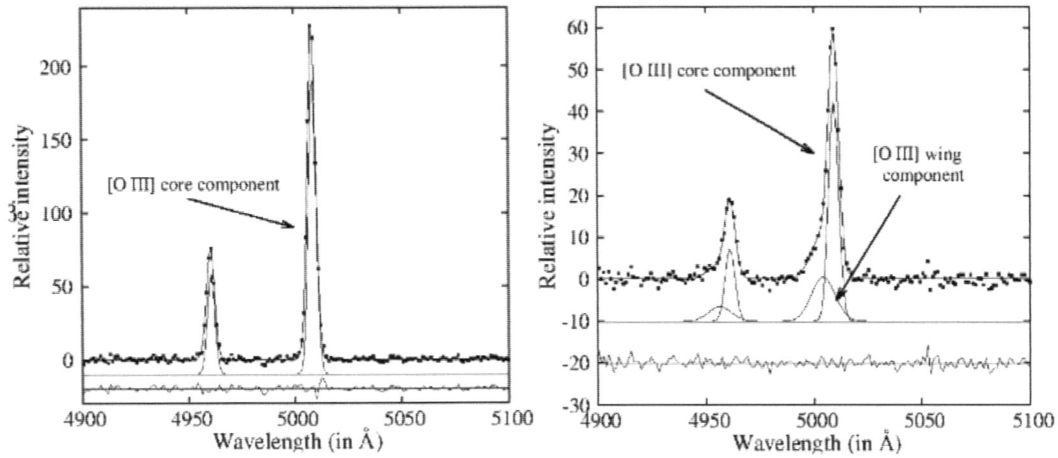

Figure 1: The example of the fit of object with no asymmetry in [O III] line profile (left), and with present asymmetry in [O III] line profile (right).

3. RESULTS AND CONCLUSIONS

The recent investigations of the NLR kinematics (Woo et al. 2016, Eun et al. 2017) suggest that, in order to avoid the outflow influence to the line width, wing component should be subtracted from the [O III] profile, and only the core component should be used as surrogate for the σ^*. Here we investigate if there is some influence of the outflow kinematics, not only in the wings, but also in the core components as well. We compared the widths of the [O III] core components with the σ^*, for both subsamles. We found that correlation between widths of [O III] core components and σ^* is significantly stronger for subsample with absent wing components ($r = 0.70$, $P = 0$), than for the subsample with strong wing components ($r = 0.29$, $P = 1.7E-6$). The correlations are shown in Fig 2a, b. Also, we compared the [O III] luminosity with the width of the [O III] core component for the both subsamples, and we colored the symbols with diverse colors for different $M_{BH}(\sigma^*)$ bin intervals (M_{BH} is estimated using M_{BH} - σ^* relation given in Woo et al. 2015). In the subsample with absent wing components, correlation between these two parameters is stronger ($r = 0.53$, $P = 4.4E-11$), and it could be seen that L [O III] and width of the [O III] core component are both indicators of the $M_{BH}(\sigma^*)$ (see Fig 3a). For subsample with present wing components, there is only weak trend between these two parameters, and they are not correlated with $M_{BH}(\sigma^*)$ (see Fig 3b).

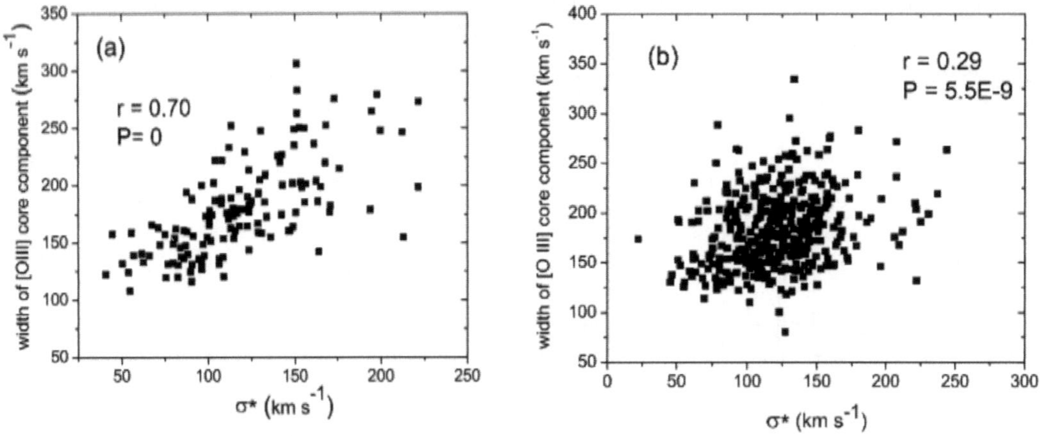

Figure 2: The correlation between σ * and width of [O III] line core component for the subsample with no asymmetry in [O III] line profile (a), and for subsample with present asymmetry in [O III] line profile (b).

As it can be seen from these correlations, the width of the [O III] core component is better surrogate for σ* in the subsample with absent outflows, and consequently better M_{BH} indicator.

This implies that in the case when outflow kinematics is strong in spectra (strong wing components in narrow emission lines), there is probably some influence of the outflow in the narrow core component as well. These results are in accordance with the results obtained from spatial spectroscopy of the several AGNs Type 2 by Karouzos et al. 2016. They found that outflows contribute, not only to the wing, but also to the [O III] core component.

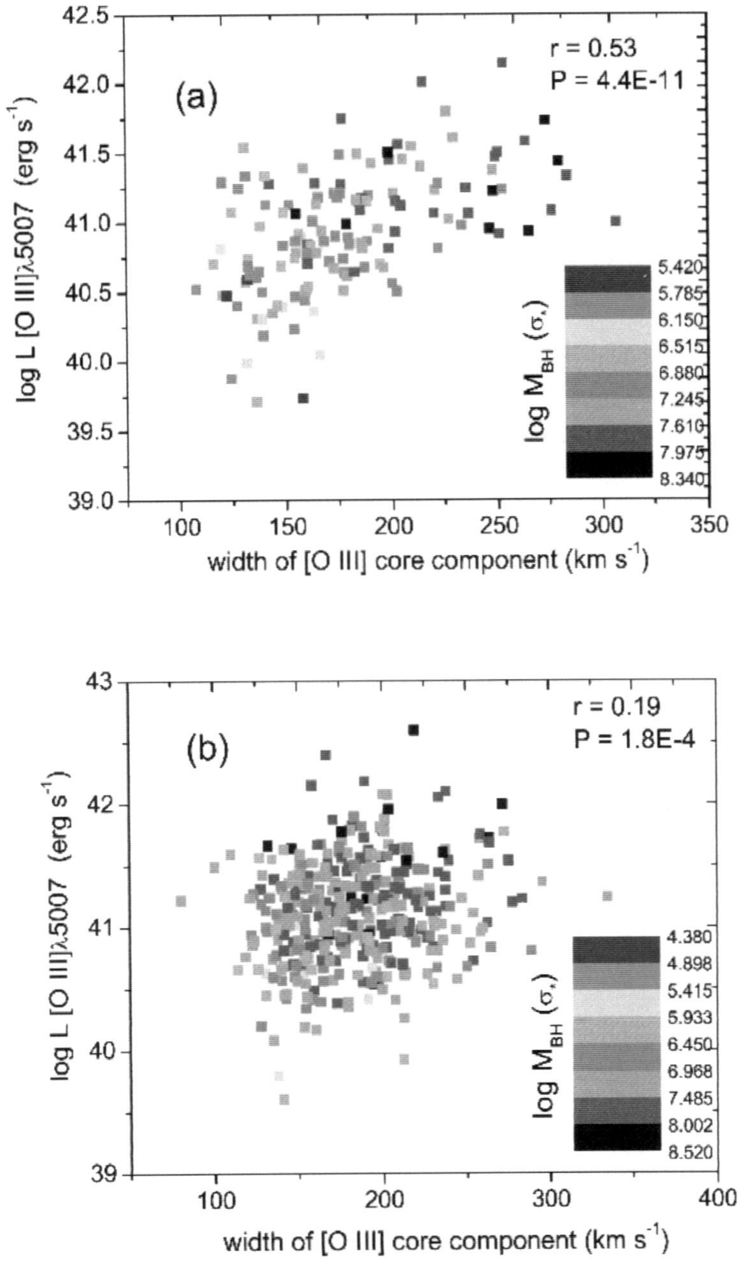

Figure 3: The correlation between L[O III] and width of [O III] line core component for the subsample with no asymmetry in [OIII] line profile (a), and for subsample with present asymmetry in [O III] line profile (b).

Our results point out that M_{BH} estimation using the width of the [O III] core as surrogate for the σ* can be applied for the AGNs Type 2 with absent wing components (with no asymmetry in emission line profiles), in order to get accurate M_{BH} estimation. In the case of the AGNs Type 2 with significant emission line asymmetry, this method should be taken with the caution because of the large scatter caused by the outflow influence.

Our future investigation will be directed to the analysis of the other emission lines in AGN type 2 spectra, in order to check if there are some emission lines which are less affected with outflow kinematics than [O III] lines, and therefore better σ* surrogate.

References

Antonucci R.: 1993, *ARA&A*, 31, 473.
Bonning E. W., Shields G. A., Salviander S., McLure R. J.: 2005, *ApJ*, 626, 89.
Eun D., Woo J.-H., Bae H.-J. 2017, *ApJ,* 842, 5.
Greene J. E., Ho L. C.: 2005, *ApJ*, 630, 122.
Grupe D., Mathur S., 2004, *ApJ*, 606, 41.
Karouzos M., Woo J.-H., Bae H.-J.: 2016, *ApJ*, 819, 148.
Lakićević M., Kovačević-Dojčinović J., Popović L. Č.: 2017, *MNRAS*, 472, 334.
Nelson C. H., Whittle M.: 1996, *ApJ*, 465, 96.
Nelson C. H.: 2000, *ApJ*, 544, 91.
Salviander S., Shields G. A.: 2013, *ApJ*, 764, 80.
Shields G. A., Gebhardt K., Salviander S. et al.: 2003, *ApJ*, 583, 124.
Tremaine S., Gebhardt K., Bender R. et al.: 2002, *ApJ*, 574, 740.
Urry C. M., Padovani P.: 1995, *PASP*, 107, 803.
Woo J.-H., Bae H.-J., Son D. and Karouzos M.: 2016, *ApJ*, 817, 108.
Woo J.-H., Yoon Y., Park S., Park D., Kim S. C.: 2015, *ApJ*, 801, 38.

Proceedings of the XII Serbian-Bulgarian Astronomical Conference
(XII SBAC) Sokobanja, Serbia, September 25-29, 2020
Editors: L. Č. Popović, V. A. Srećković, M. S. Dimitrijević and A. Kovačević
Publ. Astron. Soc. "Rudjer Bošković" No 20, 2020, 39-41

INFLATIONARY MODELS, REHEATING AND SCALAR FIELD CONDENSATE BARYOGENESIS

DANIELA KIRILOVA and MARIANA PANAYOTOVA

Institute of Astronomy and NAO, Bulgarian Academy of Sciences
Blvd. Tsarigradsko Shosse 72, Sofia, Bulgaria
E-mail: dani@astro.bas.bg, mariana@astro.bas.bg

Abstract. We discuss the baryon asymmetry value generated for 70 sets of parameters of the Scalar Field Condensate Baryogenesis (SFC) model in different inflationary scenarios and for different reheating scenarios. In previous paper we have found sets of SFC model's parameters, for which the observed value of the baryon asymmetry of the Universe can be successfully generated in the following inflationary scenarios: modified Starobinsky inflation, chaotic inflation in SUGRA and chaotic inflation in case of delayed thermalization. Here we expand our analysis to study baryon asymmetry generation in quintessential inflation. We have found several sets of parameters of SFC baryogenesis model for which successful baryogenesis is possible in case of quintessential inflation. On the contrary new inflation, Shafi-Vilenkin chaotic inflation and MSSM inflation lead to baryon asymmetry generation by several orders of magnitude higher than the observed one.

1. INTRODUCTION

In this work we present an update of our results concerning the generation of baryon asymmetry in Scalar Field Condensate (SFC) baryogenesis model (first studied in Dolgov and Kirilova (1990, 1991) in different inflationary scenarios and for different possibilities of the reheating after inflation. Preliminary results on these issues were presented in Kirilova and Panayotova (2019).

SFC baryogenesis model is among the preferred baryogenesis models today because it allows the generation of the baryon asymmetry of the Universe to proceed at lower energies, thus it is compatible with inflationary paradigm and also evades the problem of overabundance of gravitino and of magnetic monopoles. According to SFC baryogenesis model the baryon excess B is contained in baryon charge carrying scalar field φ, which transfers it to quarks during its decay. The baryon excess is generated at earlier epochs due to B-violating terms in the potential of φ.

An important feature of this model is that after inflation φ oscillates around its equilibrium point and its amplitude decreases due to the Universe expansion and the particle creation caused by the coupling of the scalar field to fermions $g\varphi f_1 f_2$, where $g^2/4\pi = \alpha_{GUT}$ (Dolgov and Kirilova 1990; Kirilova and Panayotova 2007). Hence, the baryon charge B, contained in φ condensate, is reduced due to particle production at the high energy stage, where BV is considerable. If the rate of particle creation Γ is a decreasing function of time, the damping is slow and B survives until the decay of φ, when B is transferred to fermions. The generated baryon asymmetry strongly depends on α.In this work we have accounted for the particle creation process numerically.

The produced baryon asymmetry depends on the baryon excess B, the reheating temperature of the Universe T_R and the value of the Hubble parameter at the inflationary stage H_I. Hence, there is a big variety of possible H_I and T_R values following from different inflationary and reheating scenarios.

Today there exist numerous inflationary models. Besides the reheating process at the end of the inflationary stage may have proceeded through perturbative and nonperturbative mechanisms (Traschen, Brandenberger 1990; Dolgov, Kirilova 1990; Kofman, Linde, Starobinski 1994) and also through different decay channels and different decay rates of the inflaton and different thermalization (instant or delayed). The efficient or delayed thermalization were discussed in (Mazumdar, Zaldivar 2014).

2. INFLATIONARY SCENARIOS, REHEATING AND BARYOGENESIS

We have analyzed the baryon asymmetry generation in SFC baryogenesis model for more than 70 sets of parameters of the SFC model, presented in detail in (Kirilova, Panayotova 2015; Kirilova, Panayotova 2014). The studied ranges of the models parameters were: $H_I = 10^7$ - 10^{16} GeV, m = 100 - 1000 GeV, $\alpha = 10^{-3}$ – 5.10^{-2}, $\lambda_i = 10^{-3} - 5.10^{-2}$.

Now there exist numerous models of inflation (Martin, Ringeval, Vennin, 2014). In our previous work we have discussed the following inflationary scenarios: new inflation, chaotic inflation, Shafi-Vilenkin chaotic inflation, chaotic inflation in SUGRA, Starobinsky inflation and MSSM inflation. It was found that there exist possibility for generation of the baryon asymmetry with similar to the observed value $\boldsymbol{\beta}_{obs} = 6.10^{-10}$, where $\boldsymbol{\beta} \sim BT_R/H_I$, for the following models: the simplest Shafi-Vilenkin model (Shafi, Vilenkin 1984) in chaotic inflation for $T_R = 10^{12}$ - 10^{13} GeV again in case of delayed thermalization; modified Starobinsky inflation (Starobinsky 1980) for $T_R = 10^9$ GeV, $H_I = 10^{11}$ GeV, for the efficient thermalization and chaotic inflation in SUGRA (Nanopoulos, Olive, Srednicki 1983) $T_R > 10^9$ GeV. All other considered models were shown to produce baryon asymmetry many orders of magnitude bigger than the observed one.

Here we consider quintessential inflationary models (Peebles, Vilenkin 1999), which are among the preferred inflationary models today, because they allow simultaneous explanation of the inflationary expansion at the very early universe and the accelerated expansion of the Universe at late epochs using a single scalar field potential:

$$V = \lambda(\phi^4 + M^4), \quad \phi < 0$$

$$V = \lambda M^8/(\phi^4 + M^4), \quad \phi \geq 0$$

At $\phi >> M$ this is a chaotic inflation potential, at . $\phi << M$ it is a quintessence form, $\lambda \sim 10.^{-14}$. Some model modifications were provided recently to obtain agreement with the Plank18 observational data

We have considered different types of reheating after inflation, in particular models with instantaneous and with delayed thermalization were analyzed. We have found that in quintessential inflation for $T_R = 2.10^5$ GeV, $H_I = 10^{12}$ GeV there exist several realizations of the SFC baryogenesis model with successful production of the observed baryon asymmetry value, namely for m = 350 GeV and $\alpha = 10^{-3}$ and the following range of models self-coupling parameters: $\lambda_1 = 10^{-3} - 5.10^{-2}$, $\lambda_2 = \lambda_3 = 10^{-4} - 5.10^{-3}$.

Thus, the numerical analysis of SCF baryogenesis model in quintessential inflationary model predicts successful generation of the observed baryon asymmetry of the Universe for several sets of SCF model parameters.

Planck missions releases have put constraints on the inflationary models.

It is interesting that most of the inflationary models in which successful SCB baryogenesis is possible are also among the observationally preferred ones by the latest Planck data.

Acknowledgments

We acknowledge the partial financial support by project DN18/13-12.12.2017 of the Bulgarian National Science Fund of the Bulgarian Ministry of Education and Science.

References

Dolgov A. D., Kirilova D.: 1990, *Sov. J. Nucl. Phys.*, 51, 172.

Dolgov A. D., Kirilova D.: 1991, *J. Moscow Phys. Soc.*, 1, 217.

Kirilova D., Panayotova M. 2019, *AIP Conf. Proceedings, Conference: 10th Jubilee International Conference of the Balkan Physical Union*, 2075, 090017.

Kirilova D., Panayotova M.: 2007, *Bulg. J. Phys.*, 34 s2, 330.

Traschen J. H., Brandenberger R. H.: 1990, *PRD*, 42, 2491.

Kofman L., Linde A., Starobinski A.: 1994, *Phys. Rev. Lett.*, 73, 3195.

Mazumdar A., Zaldivar B.: 2014, *Nucl. Phys. B*, 886, 312.

Kirilova D., Panayotova M.: 2015, *Advances in Astronomy*, 2015, 425342.

Kirilova D., Panayotova M.: 2014, *BAJ*, 20, 45.

Martin J., Ringeval C., Vennin V.: 2014, *Phys. Dark Univ.*, 5-6, 75.

Shafi O., Vilenkin A.: 1984, *Phys. Rev. Lett.*, 52, 691.

Starobinsky A.: 1980, *Phys. Lett. B*, 91, 99.

Nanopoulos D. V., Olive K., Srednicki M.: 1983, *Phys. Lett. B*, 127, 30.

Peebles P. J. E., Vilenkin A.: 1999, *Phys. Rev. D*, 59, 063505.

Proceedings of the XII Serbian-Bulgarian Astronomical Conference (XII SBAC) Sokobanja, Serbia, September 25-29, 2020
Editors: L. Č. Popović, V. A. Srećković, M. S. Dimitrijević and A. Kovačević
Publ. Astron. Soc. "Rudjer Bošković" No 20, 2020, 43-49

EVOLUTION OF MASSIVE BINARY SYSTEMS: PRIMARY STAR EVOLUTION INTO A NEUTRON STAR

JELENA PETROVIC

Astronomical Observatory, Volgina 7, 11000 Belgrade, Serbia
E-mail: jpetrovic@aob.rs

Abstract. The sources of gravitational waves, observed by the LIGO and Virgo detectors, have been linked with mergers in binary systems (Abbott et al. 2019). Those systems are consisting of compact objects - black holes and/or neutron stars. Progenitors of those systems are massive binary systems and initial masses of above ~30 $M_\odot$ are needed for the formation of double black holes. We present examples of evolutionary models of massive binary systems with initial masses around 30 $M_\odot$ and we follow their evolution until the primary stars form an iron core. The initial orbital period is set at 3 days and the initial mass ratio is 0.9. Those models are calculated with the MESA (Modules for Experiments in Stellar Astrophysics) stellar evolution numerical code. We find that primary stars in those systems evolve into neutron stars.

1. INTRODUCTION

Recent discovery of gravitational waves by the LIGO and Virgo collaborations (Abott et al. 2019) has linked those events with mergers in double compact binary systems, consisting of neutron stars and/or black holes, the end products of stellar evolution in massive close binary systems. Initial masses of above ~30 $M_\odot$ are shown to be necessary to produce black hole relics (Kruckow et al. 2018).

The primary star in a binary system, the component with the greater mass, evolves faster than the secondary and through envelope expansion may reach the radius of its Roche lobe and starts transferring mass onto the secondary star. This process is called Roche Lobe Overflow (RLOF). A star in a binary can reach its Roche lobe radius during different phases of evolution: core hydrogen burning, shell hydrogen burning or after the onset of central helium burning, corresponding respectively to Case A, B and C of mass transfer. However, in case there is a mass transfer during the core hydrogen burning – Case A, mass transfer during the shell hydrogen burning phase is called Case AB. Case ABB is then the third mass transfer that takes place after helium core burning is completed.

Evolutionary models of massive binary stars, evolving via the stable mass transfer channel and considering different accretion efficiencies, have been presented by many authors (for example De Greve and De Loore 1992, Pols 1994, De Loore and Vanbeveren 1994, Wellstein and Langer 1999, Wellstein et al. 2001, Petrovic et al. 2005). Wellstein et al. (2001) have shown that the initial mass ratio in massive binaries should be near unity, otherwise the system evolves into a contact, or in other words, the secondary star expands due to accretion and both stars fill their Roche lobes. Petrovic et al. (2005) have modeled progenitor evolution of observed Wolf Rayet + O binary systems an they found that the most likely evolution is via Case A mass transfer with accretion efficiency of only 10%.

Double compact objects are relics of massive binary star evolution.

Such binary systems start as double O-type stars and evolve through multiple interactions in their lifetimes, transferring matter and angular momentum from one to another. Those systems evolve through a Wolf-Rayet + O phase and survive two supernova explosions. The orbits of massive close binaries are most probably tidally circularized and eccentricity is not an important parameter to consider (Hurley 2002).

In this paper, we present the calculations of the evolution of the initial binary systems: 30 + 27 $M_{\odot}$ and 32 + 28.8 $M_{\odot}$, both with initial orbital period of 3 days and initial mass ratio of 0.9. We follow the evolution of the primary star through three Roche lobe overflows (Case A, AB and ABB) until iron core formation.

2. MODELS

For the calculation of the evolution of binary models presented in this paper, the MESA (Modules for Experiments in Stellar Astrophysics) code was used (Paxton et al. 2011, 2013, 2015, 2018). Models with initial masses 30 + 27 $M_{\odot}$ and 32 + 28.8 $M_{\odot}$, both with initial orbital period of 3 days and initial mass ratio of 0.9 are presented. Metallicity is set to 0.02.

The MESA code calculates simultaneously the evolution of both stars within a binary system. Mass transfer happens via the L_1 Lagrangian point and its rate is calculated according to the Ritter scheme (Ritter 1988). The composition of accreted material is identical to the donor's current surface composition. It is assumed that the mass lost in the stellar wind has the specific orbital angular momentum of its star. For the case of inefficient mass transfer, angular momentum loss follows Soberman et al. (1997) where fixed fractions of the transferred mass are lost as a fast isotropic wind. Stellar wind mass loss is calculated according to Vink et al. (2001).

3. RESULTS AND DISCUSSION

The modeled binaries start their evolution as detached systems with an orbital period of 3 days and evolve through three mass transfers. We follow the evolution of the binary systems until iron core formation in the primary stars.

The primary stars in both binary systems (30 + 27 $M_{\odot}$ and 32 + 28.8 $M_{\odot}$) fill their Roche lobe first time during the core hydrogen burning phase. At this time they still have about 30% of hydrogen left in their cores. During Case A mass transfer, the primary stars lose a significant amount of their mass (about 15 $M_{\odot}$). The secondary stars accrete only 10% of the mass lost by the primaries, and due to the high stellar wind mass loss, they actually slightly decrease their initial masses. After the core hydrogen burning phase, the primary stars expand, mass transfer starts again and lasts until the primary stars ignite helium in their cores. Finally, after the core helium burning phase, there is one more expansion of the primary star that leads to Roche lobe overflow.

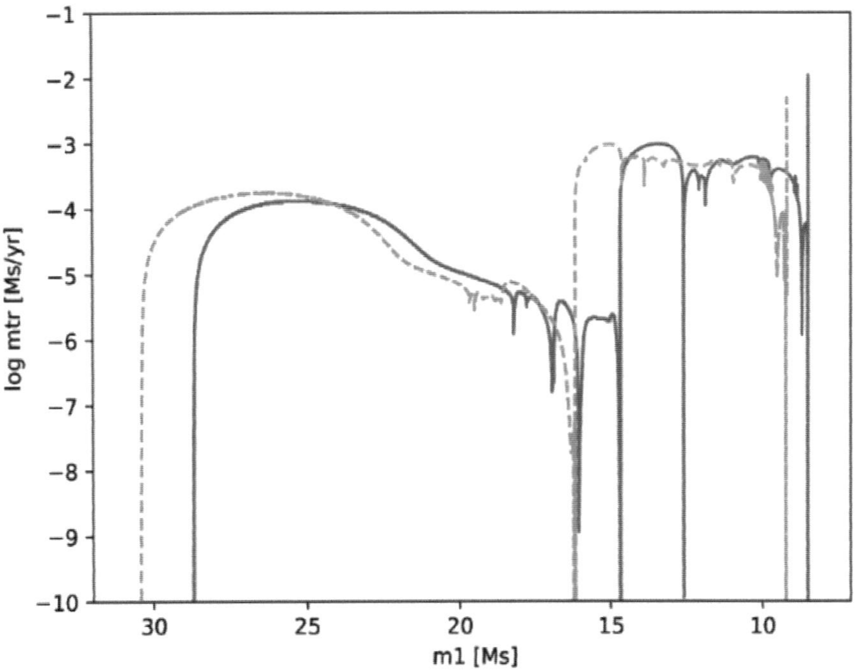

Figure 1: Mass transfer rate as a function of the primary mass in binary systems 30 + 27 $M_{\odot}$ (blue solid line) and 32 + 28.8 $M_{\odot}$ (orange dashed line), both with initial orbital period of 3 days and accretion efficiency of 10%.

Figure 1 shows the mass transfer rate in modeled systems as a function of the primary mass. The first mass transfer (Case A) starts when the primary stars are 28.8 and 30.5 M⊙, slightly less massive than initially due to the stellar wind mass loss. After Case A mass transfer, the primary masses are 14.7 and 16.2 M⊙. The corresponding secondary masses are 26.9 and 28.4 M⊙. When the core helium burning phase starts, the primary masses are 8.4 and 9.2 M⊙ and the secondary masses are 27.4 and 29.0 M⊙.

The mass transfer rate during Case A in both systems is in the order of magnitude of 10^{-4} M⊙/yr in its maximum. During the following mass transfer, it reaches a value one order of magnitude higher. The last peak on this plot represents the beginning of the last mass transfer. At the same time, an iron core is formed in the primary stars.

During the evolution from the main sequence to the formation of the iron core, the orbital period increases from the initial 3 days to 7.7 and 7.6 days during Case A mass transfer and to 27.1 and 26.8 days during the subsequent mass transfer.

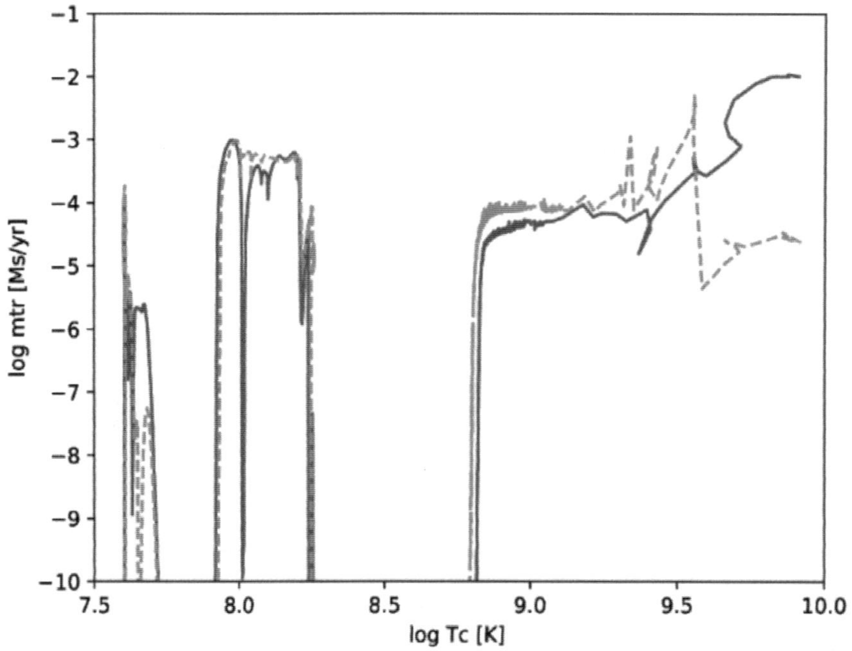

Figure 2: Mass transfer rate as a function of the primary star central temperature in binary systems 30 + 27 M⊙ (blue solid line) and 32 + 28.8 M⊙ (orange dashed line), both with initial orbital period of 3 days and accretion efficiency of 10%.

Figure 2 also shows mass transfer rate, but as a function of the central temperature of the primary star. Here we can clearly see three mass transfer phases for both systems. The last mass transfer starts after helium core burning is completed and it corresponds with a fast increase of the central temperature accompanied by the formation of a carbon, oxygen, silicon and iron core in the primary stars in the modeled systems.

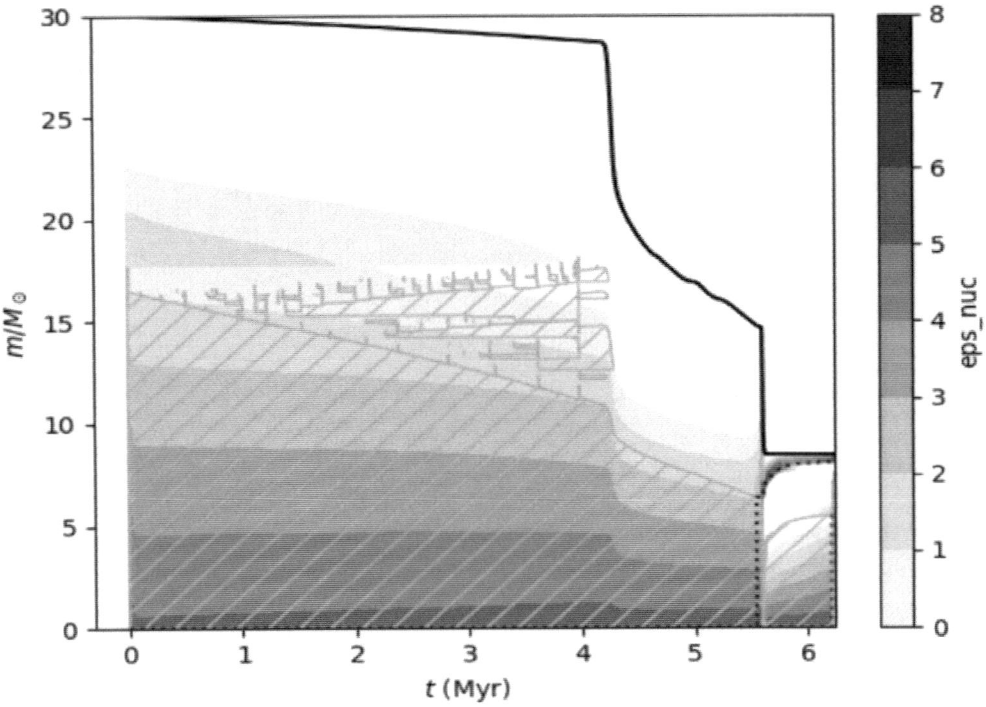

Figure 3: Convective plot of the primary star in binary system 30 + 27 M⊙ for initial orbital period of 3 days and accretion efficiency of 10%. X-axis shows time and y-axis shows stellar mass. Top black line presents the mass of the primary. Blue regions mark nuclear burning zones, darker shades indicate large intensity. Green diagonally hatched areas indicate convection zones. Blue dotted line presents mass of helium core and red dotted line mass of carbon-oxygen core.

Figure 3 shows a so-called Kippenhahn plot of the internal structure evolution of the primary star in the system 30 + 27 M⊙. The top black line presents the mass of the primary. At the age of about 4 million years and 5.6 million years, large mass loss due to mass transfers is visible. The helium core is formed at about 5.6 million yeas and its mass is 8.14 M⊙, the carbon-oxygen core is formed at about 6.2 million years with a mass of 5.41 M⊙.

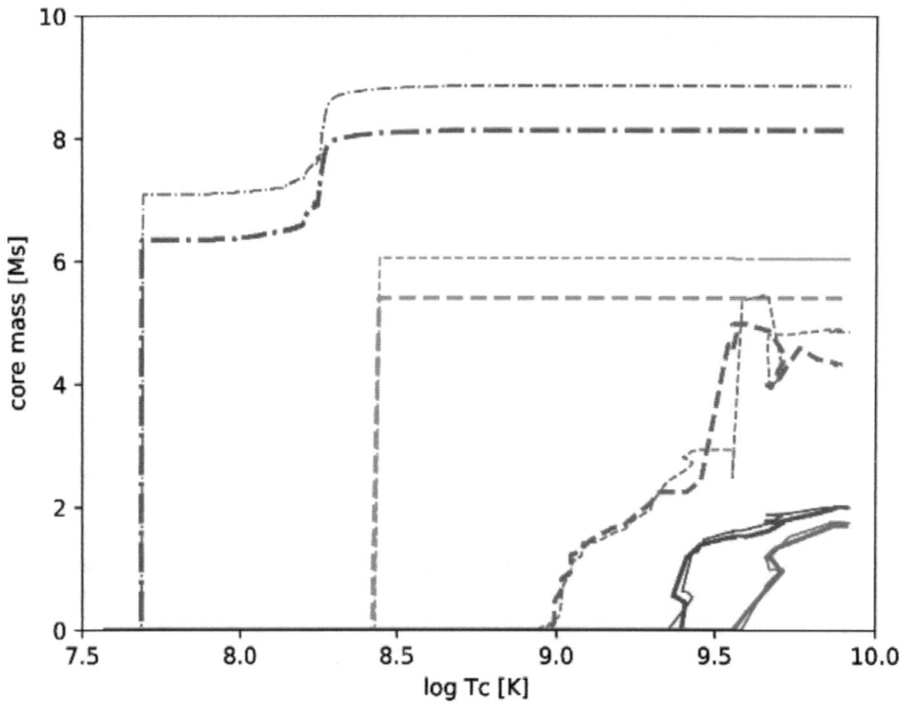

Figure 4: Core masses as a function of central temperature for primary stars in binary systems 30 + 27 M⊙ (thick lines) and 32 + 28.8 M⊙ (thin lines) with initial period of 3 days and accretion efficiency of 10%. Helium core mass is presented with a blue line, carbon with orange, oxygen with green, silicon with red and iron with a purple line.

In binary system 32 + 28.8 M⊙, the masses of the helium and CO cores are slightly larger due to the larger initial mass of the primary. Figure 4 shows helium, carbon, oxygen, silicon and iron core masses as a function of central temperature in primary stars for both binary systems. The mass of the helium and carbon-oxygen core in binary system 32 + 28.8 M⊙ are 8.87 and 6.04 M⊙ respectively. Both presented primary stars have a final carbon-oxygen core less massive than 6.5 M⊙, which indicates that they evolve into neutron stars after an iron-core collapsing supernova explosion (Kruckow et al. 2018, Tauris et al. 2015).

4. CONCLUSIONS

In this paper, we present evolutionary paths of primary stars in two massive binary systems: 30 + 27 M⊙ and 32 + 28.8 M⊙ with initial orbital period of 3 days and accretion efficiency of 10%. Initial parameters and accretion efficiency value are selected based on the progenitor models of Wolf-Rayet + O observed binary systems by Petrovic at al. (2005) and evolutionary models by Wellstein et al. (2001).

The presented calculations follow the evolution of two massive binary systems until the primary forms an iron core. Subsequently, the primary will explode as a supernova and leave a neutron star as a relic (CO core mass < 6.5 $M_{\odot}$). If the binary system remains bound after this SN explosion (type Ib/c), it will eventually be observable as a high-mass X-ray binary (HMXB).

After some time, the secondary star expands enough to fill its Roche lobe and starts transferring mass onto the neutron star. The system may become dynamically unstable and enter the so-called common envelope (CE) phase. In this phase, dynamical friction of a neutron star inside an envelope of a secondary star results in a large orbital momentum loss and a decrease in orbital period. Eventually, the secondary star also explodes as a supernova, leaving a compact object as a relic.

In the cases of the calculated systems, at the time of the supernova explosions of the primary stars, secondaries, that are still core hydrogen burning stars, have masses of about 27 and 29 $M_{\odot}$. Those masses indicate that the presented systems likely evolve into double neutron stars. If their orbital periods stay relatively short after the second supernova explosion, such double compact binary systems eventually merge and give rise to powerful emission of gravitational waves.

References

Abbott B. P. et al. (LIGO Scientific Collaboration and Virgo Collaboration): 2019, *Phys. Rev. X*, 9, 031040.

De Greve J. P., De Loore C.: 1992, *A&AS*, 96, 653.

De Loore C., Vanbeveren D.: 1994, *A&A*, 292, 463.

Hurley J. R., Tout, C. A., Pols O. R.: 2002, *MNRAS*, 329, 897

Kruckow M. U., Tauris M. T., Langer N. et al.: 2018, *MNRAS*, 481, 1098.

Paxton B., Bildsten L., Dotter A. et al.: 2011, *ApJS*, 192, 3.

Paxton B., Cantiello M., Arras P. et al.: 2013, *ApJS*, 208, 4.

Paxton B., Marchant P., Schwab J. et al.: 2015, *ApJS*, 220, 15.

Paxton B., Schwab J., Bauer E. B. et al.: 2018, *ApJS*, 234, 34.

Petrovic J., Langer N., van der Hucht K.: 2005, *A&A*, 435, 1013.

Pols O.: 1994, *A&A*, 290,119.

Ritter H.: 1988, *A&A*, 202, 93.

Soberman G. E., Phinney E. S., van den Heuvel E. P. J.: 1997, *A&A*, 327, 620.

Vink J.S., de Koter, A. Lamers H. J. G. L. M.: 2001, *A&A*, 369, 574.

Tauris T. M., Langer N., Podsiadlowski P.: 2015, *MNRAS*, 451, 2123.

Wellstein S., Langer N.: 1999, *A&A*, 350, 148.

Wellstein S., Langer N., Braun H.: 2001, *A&A*, 369, 939.

**Proceedings of the XII Serbian-Bulgarian Astronomical Conference
(XII SBAC) Sokobanja, Serbia, September 25-29, 2020
Editors: L. Č. Popović, V. A. Srećković, M. S. Dimitrijević and A. Kovačević
Publ. Astron. Soc. "Rudjer Bošković" No 20, 2020, 51-59**

ANALYSIS OF THE DENSITY DISTRIBUTION IN STAR-FORMING CLOUDS: EXTRACTION OF A SECOND POWER-LAW TAIL

LYUBOV MARINKOVA[1], TODOR V. VELTCHEV[1] and SAVA DONKOV[2]

[1]*Department of Astronomy, University of Sofia, James Bourchier Blvd. 5, Sofia 1164, Bulgaria*
[2]*Department of Applied Physics, Faculty of Applied Mathematics, Technical University of Sofia, 8 Kliment Ohridski Blvd., BG-1000, Sofia, Bulgaria*
E-mail: ln@phys.uni-sofia.bg, eirene@phys.uni-sofia.bg, savadd@tu-sofia.bg

Abstract. Clues to understand physics and evolution of molecular clouds can be provided through analysis of the probability density functions of mass density (ρ-pdf) and of column density (N-pdf). Many numerical simulations show that a power-law tail (PLT) emerges at the high-density end of the ρ-pdf at advanced evolutionary stages of star-forming clouds. Later, at the stage of collapse of first formed protostellar cores, a second, shallower PLT appears (Kritsuk et al. 2011). Double PLTs have been also detected in N-pdfs from *Herschel* maps in several star-forming regions (Schneider et al. 2015, 2020).

However, it is difficult to estimate the parameters of the second PLT due to resolution constraints. We propose a technique for extraction of a second PLT in ρ/N-pdfs which is an extension of the method of Veltchev et al. (2019) for extraction of single PLTs from arbitrary density distributions. The technique is applied to a set of hydrodynamical simulations of isothermal self-gravitating clouds. The results confirm the emergence of a shallower second PLT in ρ-pdfs at timescales, comparable with the free-fall time of the average density in the box. Second PLTs are detected also in N-pdfs derived from *Herschel* maps of a low-mass (Pipe) and high-mass (M 17) star-forming regions.

1. INTRODUCTION

Stars are born in molecular clouds (MCs), therefore the study of star formation requires understanding of the morphological and kinematical evolution of MCs. Initial stage of cloud formation is the compression of interstellar warm atomic gas by supersonic flows followed by rapid cool down due to non-linear thermal instabilities (see Ballesteros-Paredes et al. 2020, for a review). Stars begin to form when self-gravity in the cloud takes slowly over and local sites of gravitational collapse emerge. This evolution could be described in terms of detection and physical parameters of substructures (clumps, cores) or in terms of indicators of general structure.

An important indicator of general cloud structure is the probability distribution function (pdf) of mass density (ρ-pdf) and its analysis can give clues to understand the physics and evolution of the cloud. From observations, one could derive the pdf of column density (N-pdf), which turns out to be morphologically analogous to the ρ-pdf. In isothermal, non-gravitating fluids with well developed supersonic turbulence the ρ-pdf is mostly lognormal (e.g. Vázquez-Semadeni 1994, Li, Klessen and Mac Low 2003, Federrath et al. 2010), i.e. can be fitted by lognormal function of type:

$$p(s)\, ds = \frac{1}{\sqrt{2\pi\sigma^2}} \exp\left[-\frac{1}{2}\left(\frac{s - s_{\max}}{\sigma}\right)^2 \right] ds \,,$$

where $s=\log(\rho/\rho_0)$ and $s_{\max}$ are the logdensity (with normalization to the mean density ρ_0) and its value at the distribution peak and σ is the standard deviation. This result was confirmed by numerous simulations. At advanced evolutionary stages, when self-gravity becomes important in the energy balance in the cloud, a power-law tail (PLT) with functional form

$$p(s) = A \exp(qs) = A(\rho/\rho_0)^q \,, \quad s \geq s_{\rm PLT} \,,$$

emerges at the high-density end of the pdf, where A is a constant, q is the power index and the deviation point (DP) $s_{\rm PLT}$ from lognormality separates the two regimes (Klessen 2000, Kritsuk, Norman and Wagner 2011, Federrath and Klessen 2013). The evolution of the N-pdf turns out to be morphologically similar (see, e.g, Ballesteros-Paredes et al. 2011, Koertgen, Federrath and Banerjee 2019). Example of a pdf with main lognormal part and a PLT is shown in Fig. 1.

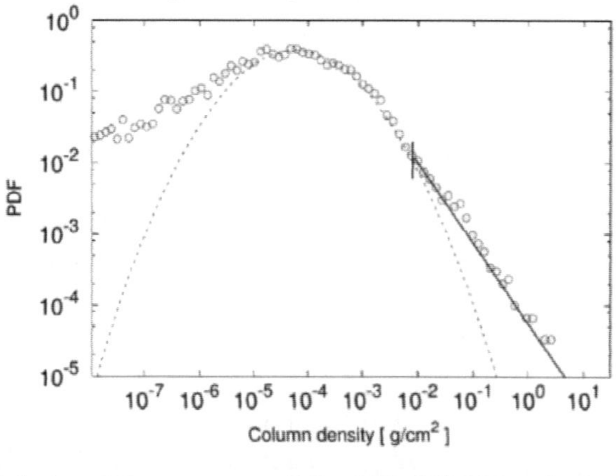

Figure 1: N-pdf derived from a run of the SILCC (SImulating the LifeCycle of molecular Clouds) simulations (Girichidis et al. 2018).

In the course of further MC evolution the main part of the ρ-pdf retains its (quasi-) lognormal shape. On the other hand, the slope of the PLT gets slowly shallower, tending toward a constant value while DP shifts to lower values.

Some numerical studies with high resolution (reaching down AU scales) hint at emergence of a *second* PLT at advanced evolutionary stages of star-forming clouds. For instance, Kritsuk et al. (2011) found that the density distribution in self-gravitating clouds develops an extended PLT with a slope of about −1.7 at high densities on top of the usual lognormal. The tail departs from the initial lognormal distribution already at $\rho/\rho_0 \sim 10$ and continues straight for nearly 10 dex in *p(s)* and more than 6 dex in density. As the simulation progresses, the slope continues to evolve slowly toward shallower values reaching q=−1.67 at the end of the simulation. An even shallower tail is detected at densities $\rho/\rho_0 \geq 10^7$. This might indicate mass pile-up due to an additional support against gravity due to conservation of angular momentum (rotation of prestellar cores), strong magnetic fields in the densest parts of MC, change in the equation of state (non-isothermality; see Donkov et al. in this issue) or all these factors.

2. EXTRACTION OF POWER-LAW TAILS OF THE PDF

There is a methodological problem with the extraction of the PLT: its characteristics and those of the lognormal fitting function are obviously interdependent. Let us review the usual procedure to extract the PLT: (i) Find the best lognormal fit of the main pdf part (e.g. using the χ^2 goodness); (ii) Estimate the DP of the distribution from the lognormal fit (e.g. using the 3σ criterion, where σ is the Poissonian data uncertainty in the considered bin). (iii) Fit the rest of the distribution with a PL function. Such an approach rests on the assumption that the main pdf part is lognormal and thus the resulting DP and the PL slope depend on the parameters of the lognormal fit. However, if the PL regime is to be interpreted as a signature of the impact of self-gravity, then the slope value is an indicator of the cloud's evolutionary stage. We need a method to extract the PLT on minimal assumptions about the rest of the density distribution.

Such approach was recently proposed by Veltchev et al. (2019; hereafter V19) and named adapted BPLFIT method. The power-law fit of a distribution (or part of it) is derived by use of Kolmogorov-Smirnov (KS) goodness-of-fit statistics. The procedure does not rule out that other, non-power-law, functions might better fit the observed distribution – it simply derives the range and the slope of the best possible power-law fit. This method can deal with large datasets of size $< 10^5$ points from numerical simulations and high-resolution imaging of MCs and is applicable to linear, logarithmic and arbitrary binning schemes. Average slope and DP are derived as the number of bins is varied and are not sensitive to spikes and other local features of the distribution's tail.

3. TECHNIQUE FOR EXTRACTION OF A SECOND POWER-LAW TAIL

The adapted BPLFIT method can be elaborated further for detection of a second PLT (if present). The PLFIT procedure searches for the PLT of the considered PDF by use of the KS statistic for given lower cutoff $x_{\rm min}$:

$$D = \max_{x_i \geq x_{\min}} |S(x_i) - P(x_i)|$$

where $S(x_i)$ is the cumulative distribution function (CDF) of the data and $P(x)$ is the CDF of the best-fitting power-law model in the range $x_i \geq x_{\min}$. The value of $x_i \geq x_{\min}$ which minimizes D and the corresponding power-law index are selected as DP and slope of the PLT, respectively. If no lower cutoff is introduced, $x_{\min}$ is simply the lower limit of the data set (in our case, the minimal logdensity) – V19 extracted PLTs from numerical and observational PDFs in this way. Gradual increase of $x_{\min}$ constrains the considered data set and, hence, the set of values $|S(x_i) - P(x_i)|$ to obtain the KS statistic. In particular, such approach may help to detect a second PLT corresponding to higher logdensities, for some $x_{\min}$ which exceeds the DP of the single (first) PLT.

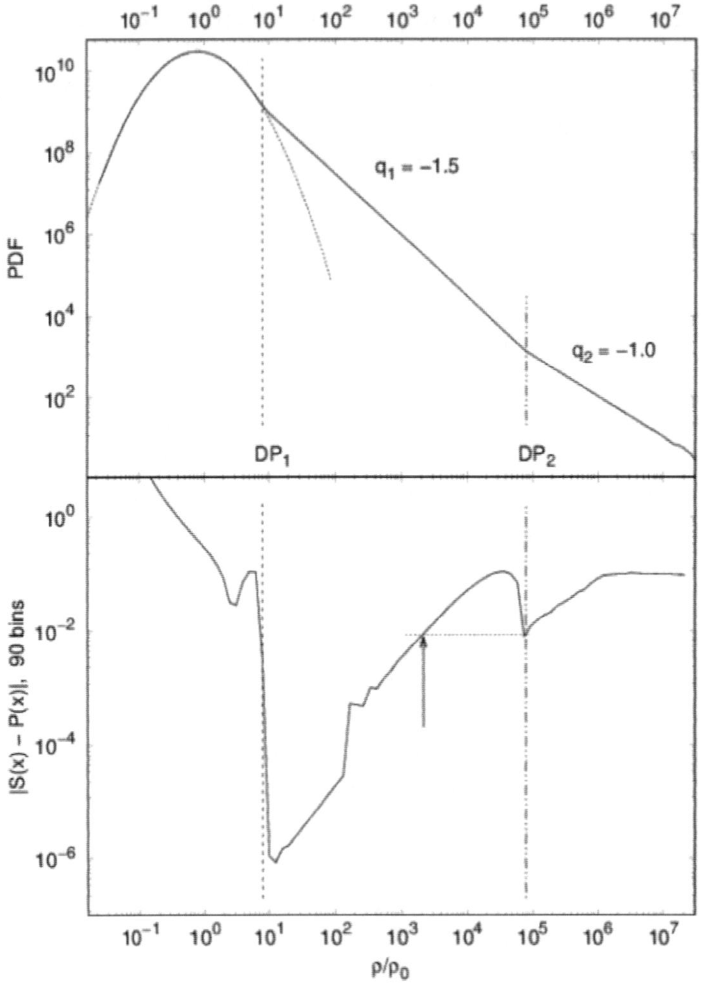

Figure 2: Illustration of the suggested method for extraction of two PLTs as applied to an analytic binned PDF (top panel; solid) with main part fitted by lognormal function (dotted) and two PLTs.
Bottom panel displays the function $|S(x) - P(x)|$ (see text).

To illustrate this we construct an analytic PDF (Fig. 2, top) whose shape and parameters resemble the one obtained in the numerical study of Kritsuk et al. (2011). The main part is lognormal while the high-density one consists of two PLTs with deviation points DP_1 and DP_2 and slopes $q_1 = -1.5$ (typical for evolved self-gravitating clouds, Girichidis et al. 2014) and $q_2 = -1$.

An example of the function $|S(x) - P(x)|$ for a large total number of bins (i.e. small bin size) is shown in Fig. 2, bottom. As expected, the value of $|S(x) - P(x)|$ is large in the range $x < DP_1$ which defines the lognormal part of the PDF. The deviation points of the two PLTs correspond to pronounced local minima, with a local maximum located in between. As long as $x_{min} < DP_1$, the adapted BPLFIT will extract a single PLT with DP = DP_1 and slope q_1 (Fig. 3). Choices of lower cutoffs x_{min} DP_1 still yield a single PLT with gradually changing parameters. The second PLT with DP = DP_2 will be detected at a cutoff with $|S(x_{min}) - P(x_{min})|$ $|S(DP_2) - P(DP_2)|$ (arrow and dotted line in Fig. 2, bottom) – then the procedure selects $x_i = DP_2$ (cf. Fig. 3) since with this choice the local maximum at $\rho/\rho_0 \approx 4 \times 10^4$ is excluded.

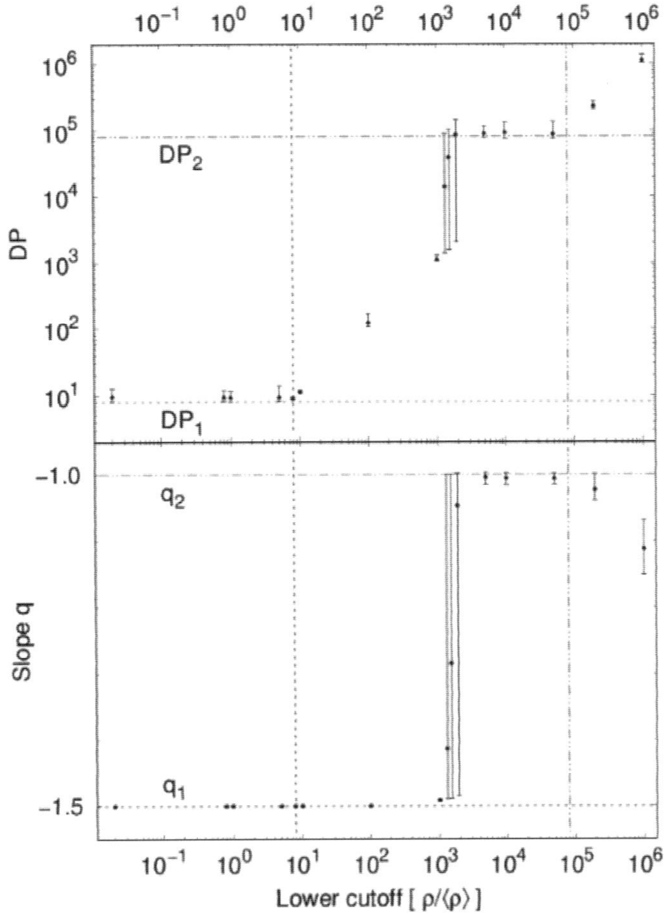

Figure 3: Dependence of the extracted PLT parameters on the chosen lower cutoff of the tested PDF (cf. Fig. 2).

In the next two Sections we present tests of the suggested method to numerical and observational data.

4. RESULTS FROM NUMERICAL DATA: DENSITY PDF

We use data from a set of 6 hydrodynamical simulations of self-gravitating clouds called HRIGT (High-Resolution Isothermal Gravo-Turbulent). The size of the numerical box is 0.5 pc, about the scale of typical large clumps in MCs. The complexity in physical modeling is reduced in favor of higher resolution and significantly higher adaptive refinement (from 256^3 up to 32768^3 cells) – thus the resolution can reach ~3 AU in high-density zones. The gas is isothermal (T=10 K) and uniformly distributed at the initial point in time. The total mass in the box is chosen to be 85 and 426 $M_\odot$ in the different runs which corresponds to 32 and 354 Jeans masses ($M_{J,0}$). The initial turbulent velocities are constructed in Fourier space with a peak of the power spectrum at k = 2, i.e. half of the box size. We distinguish between purely compressive, purely solenoidal and naturally mixed velocities (Federrath, Klessen and Schmidt 2008).

The chosen HRIGT runs differ in total mass, realizations of velocity field, turbulent driving and duration in units of free-fall time t_{ff}. In general, the runs with total mass of 354 $M_{J,0}$ have been stopped at earlier points in time (<1.5 t_{ff}). Therefore one would expect that in those cases the extracted PLTs in the mass-density pdf will be steeper. Fig. 4 demonstrates that this is indeed the case in regard to the *first* PLT. The obtained slopes are in a good agreement with the theoretical limit q~−1.5 for evolved self-gravitating clouds substantiated by Girichidis et al. (2014).

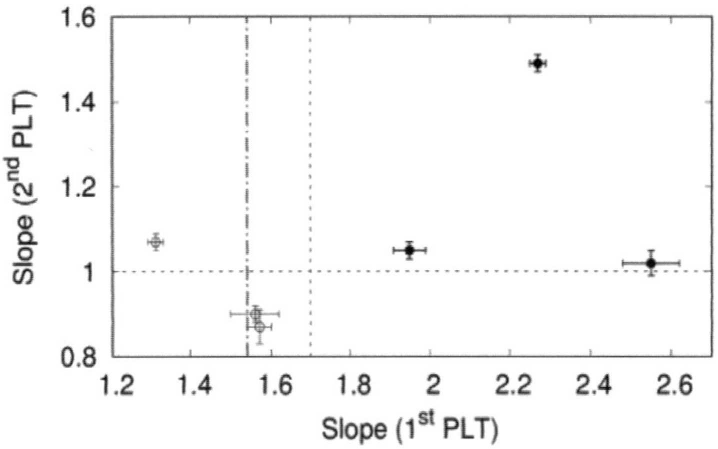

Figure 4: Comparison between the slopes of the extracted first and second PLT from the HRIGT simulations: with 32 (open circles) and 354 (filled circles) Jeans masses. The corresponding final values from Kritsuk et al. (2011; dashed) and the limiting value from Girichidis et al. (2014; dash-dotted) are plotted.

In regard to the *second* PLT, all HRIGT runs yield similar results, independent on the total mass in the box. The slopes are around -1 (and even shallower) which confirms the result of Kritsuk et al. (2011).

5. RESULTS FROM OBSERVATIONAL DATA: COLUMN-DENSITY PDF

We test the method also to N-pdfs from *Herschel* observations of several star-forming regions. The results for two of them are shown in Fig. 4. The original maps of dust emission were obtained at four wavelengths with the instruments PACS and SPIRE: 160, 250, 350, and 500 μm (see Schneider et al. 2010, 2012, for details) and convolved to a common angular resolution of 36 arcsec.

The PLT parameters of the *first* slope are consistent with the results from other numerical studies. On the other hand, N-pdfs derived from observations of regions with star-forming activity display pronounced PLTs of slopes $-2 \geq n \geq -4$ (Schneider et al. 2013, 2015a; Pokhrel et al. 2016), also in agreement with our results on the PLT evolution from the HRIGT runs.

The slope n of the N-pdf should be related to q as:

$$n = 2q/(3+q)\ ,$$

assuming that that the general cloud structure can be described through a power-law density profile (see Donkov, Veltchev and Klessen 2017, and the references therein). Plugging $q{\sim}{-1.5}$ (Girichidis et al. 2014) typical for advanced evolutionary stages in the formula above we get slopes n_2 of the *second* PLT in general agreement with the extracted ones from *Herschel* maps (Figs. 5 and 6). We conclude that the adapted BPLFIT method extracts PLTs of the ρ-pdfs and N-pdfs with slopes which are mutually consistent.

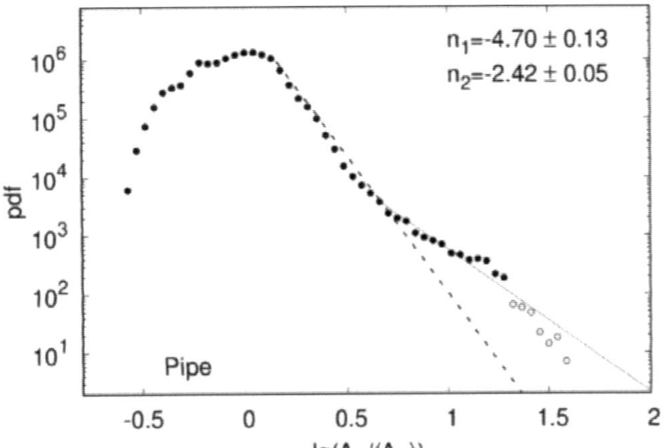

Figure 5: N-pdf with two PLTs from a *Herschel* map of the low-mass Galactic star-forming region Pipe. Open circles denote bins which were excluded from consideration due to the poor statistics.

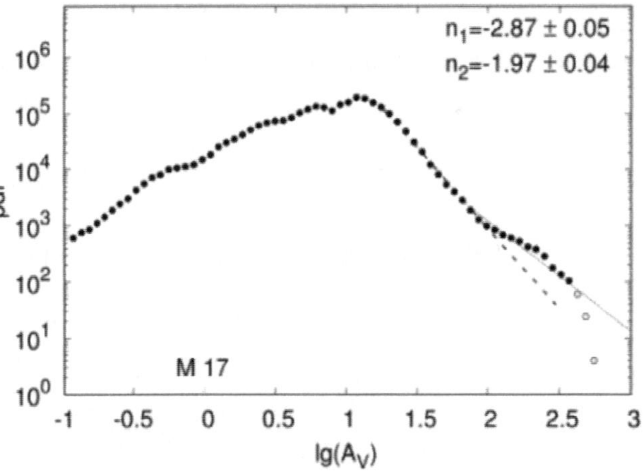

Figure 6: The same like Fig. 5 but for the high-mass Galactic star-forming region M17.

6. CONCLUSIONS

We present a novel approach for extraction of second power-law tails (PLTs) of the density (ρ-pdf) or column-density distribution (N-pdf) in star-forming clouds. The method is an extension of the adapted BPLFIT technique and was tested on data from numerical simulations of star-forming clouds at clump scale (0.5 pc; self-gravitating isothermal medium) and on observational data from *Herschel*. Our conclusions are as follows:

- The adapted BPLFIT method can be successfully extended to detect a second PLT.
- The test of this approach on numerical data with high resolution (HRIGT) yields PLT parameters in agreement with theoretical and numerical studies (Girichidis et al. 2014, Kritsuk, Norman and Wagner 2011).
- The application of the method on N-pdfs from *Herschel* data indicates the existence of a *second* PLT in N-pdfs in a dozen star-forming regions of different mass. (Schneider et al. 2020).
- A thorough comparison between the output of the methods from ρ-pdfs and N-pdfs from numerical data would shed light on the relationship between the slopes of the extracted PLTs.

Acknowledgements

The authors thank P. Girichidis for providing data from his HRIGT simulations and N. Schneider for the *Herschel* maps of the star-forming regions Pipe and M17.

L. Marinkova thanks to the Bulgarian National Science Fund for providing support through Grant KP-06-PM-38/6 (Fundamental research by young scientists and postdocs 2019). T. Veltchev acknowledges funding from the Ministry of Education and Science of the Republic of Bulgaria, National RI Roadmap Project DO1-277/16.12.2019.

References

Ballesteros-Paredes J., Andre P., Hennebelle P., Klessen R. S., Kruijssen J. M. D., Chevance M., Nakamura F., Adamo A., Vazquez-Semadeni E.: 2020, *Space Science Reviews,* 216, 5, 76.

Ballesteros-Paredes J., Vázquez-Semadeni E., Gazol A., Hartmann L., Heitsch F., Colin P.: 2011, *Monthly Notices of the Royal Astronomical Society*, 416, 1436.

Donkov S.,Veltchev T., Klessen R.S.: 2017, *Monthly Notices of the Royal Astronomical Society,* 466, 914.

Federrath C., Klessen R. S.: 2013, *The Astrophysical Journal*, 763, 51

Federrath C., Klessen R.S., Schmidt W.: 2008, *The Astrophysical Journal*, 688, L79.

Federrath C., Roman-Duval J., Klessen R. S., Schmidt W., Mac Low M.: 2010, *Astronomy & Astrophysics*, 512, A81.

Girichidis P., Konstandin L., Whitworth A. P., Klessen R. S., 2014: *The Astrophysical Journal* , 781, 91.

Girichidis P., Seifried D., Naab T., Peters T., Walch S., Wuensch R., Glover S. C. O., Klessen R. S.: 2018, *Monthly Notices of the Royal Astronomical Society*, 480, 3511.

Klessen R. S.: 2000, *The Astrophysical Journal*, 535, 869.

Koertgen B., Federrath C., Banerjee R.: 2019, *Monthly Notices of the Royal Astronomical Society*, 482, 5233.

Kritsuk A., Norman M., Wagner R.: 2011, *The Astrophysical Journal*, 727, L20.

Li Y., Klessen R., Mac Low M.-M.: 2003, *The Astrophysical Journal*, 592, 975.

Pokhrel R., Gutermuth R., Ali B., Megeath T., et al.: 2016, *Monthly Notices of the Royal Astronomical Society,* 461, 22.

Schneider N., Motte F., Bontemps S., Hennemann M., et al.: 2010, *Astronomy and Astrophysics,* 518, L83.

Schneider N., Csengeri T., Hennemann M., Motte F., et al.: 2012, *Astronomy & Astrophysics*, 540, L11.

Schneider N., Andre Ph., Konyves V., Bontemps S., et al.: 2013, *The Astrophysical Journal,* 766, L17.

Schneider N., Bontemps S., Girichidis P., Rayner T., et al.: 2015, *Monthly Notices of the Royal Astronomical Society,* 453, L41.

Schneider N., Ossenkopf-Okada V., Klessen R.S., Veltchev T., et al.: 2020, *Astronomy & Astrophysics* (submitted).

Vázquez-Semadeni E.: 1994, *The Astrophysical Journal*, 423, 681.

Veltchev T. V., Girichidis P., Donkov S., Schneider N., Stanchev O., Marinkova L., Seifried D., Klessen R. S.: 2019, *Monthly Notices of the Royal Astronomical Society*, 489, 788-801.

Proceedings of the XII Serbian-Bulgarian Astronomical Conference
(XII SBAC) Sokobanja, Serbia, September 25-29, 2020
Editors: L. Č. Popović, V. A. Srećković, M. S. Dimitrijević and A. Kovačević
Publ. Astron. Soc. "Rudjer Bošković" No 20, 2020, 61-68

STUDY OF THE FRACTAL DIMENSIONS IN THE MOLECULAR CLOUD ROSETTE BY USE OF DENDROGRAM ANALYSIS

MARIYANA BOGDANOVA, ORLIN STANCHEV and TODOR V. VELTCHEV

Department of Astronomy, University of Sofia, James Bourchier Blvd. 5, Sofia 1164, Bulgaria
E-mail: mpetrova@phys.uni-sofia.bg, o_stanchev@phys.uni-sofia.bg, eirene@phys.uni-sofia.bg

Abstract. The observed hierarchical structure in star-forming regions can be traced in terms of the fractal dimension. The latter is often defined as D=log N / log L where N is the number of fragments at given level and L is the scaling factor to the upper level. Such approach requires appropriate clump-extraction technique. An alternative approach is to explore the power-law exponent D_M of the mass-size relationship $\log(M)\propto D_M \log(L)$ where the scales L are defined in an abstract way (Beattie et al. 2019). We propose a method to derive this mass dimension D_M by use of the clump extraction technique DENDROGRAM (Rosolowsky et al. 2008). The method is applied to samples of dendrogram objects from integrated-intensity maps of ^{12}CO and ^{13}CO emissions and to dust-emission *(Herschel)* maps of the molecular cloud Rosette. The obtained scaling dependence of D_M is in general agreement with the numerical study of Beattie et al. (2019) for typical Mach numbers in molecular clouds and hints at the multifractal structure of Rosette.

1. INTRODUCTION

Observational studies of molecular clouds (MCs) in star-forming regions reveal complex sets of substructures: sheets, clumps, filaments and cores. Analysis of high-resolution maps shows that the filaments often contain dense prestellar cores of size ~0.1 pc (Andre et al. 2014) while most of the larger clumps are further decomposed to embedded condensations (see Bergin and Tafalla 2007, for review). This hierarchical, fractal cloud structure is crucial for understanding of the star formation process. Often used approaches to quantify it are studies of relationships mass *vs.* size and velocity dispersion *vs.* size in a power-law form interplay (Larson 1981, Heyer and Brunt 2004, Heyer et al. 2009) or derivation of the fractal dimension in the cloud (Elmegreen and Falgarone 1996, Elmegreen 1997).

The choice of clump-finding method may influence significantly the analysis of fractal structure – whether clumps are considered as a set of independent objects or as a hierarchy in the position-position (PP) and/or in the position-position-velocity (PPV) space. In the latter case, clump properties can be linked to the general physics of star-forming regions. A widely used hierarchical clump-finding method is the DENDROGRAM technique (Rosolowsky et al. 2008) which is appropriate for study of the fractal structure of MCs.

In this report we present a method for derivation of fractal dimensions in MCs based on their dendrogram structure derived from integrated-intensity and column-density (PP) maps. The object chosen to test the method is the Rosette MC.

2. OBSERVATIONAL DATA ON THE MOLECULAR CLOUD ROSETTE

We make use of $^{12}CO/^{13}CO$ maps (PPV cubes) taken with the 14 m telescope of Five College Radio Astronomy Observatory (FCRAO), presented and discussed by Heyer, Williams and Brunt (2006). Angular resolution of 46 arcsec allows for study of structures with sizes greater than ~0.15 pc (adopting distance to Rosette MC of 1.33 kpc, Lombardi, Alves and Lada 2011).

From the row PPV cubes we construct PP integrated-intensity maps. The process contains extraction of the channels that contain only noise, calculation of the noise levels for the rest of the channels and integration over the V axis for them. The maps of dust emission were obtained from *Herschel* observations at four wavelengths of PACS and SPIRE: 160, 250, 350, and 500 μm (see Schneider et al. 2010, 2012, for details) and convolved to a common angular resolution of 36 arcsec.

The star-forming Rosette is appropriate as a test object because of its intensive investigation in the last decades. Its local structure have been studied using various algorithms and tracers (Williams et al. 1995; Schneider et al. 1998; Dent et al. 2009; Di Francesco et al. 2010; Veltchev et al. 2018).

3. SELECTED SAMPLES OF THE DENDROGRAM METHOD

The DENDROGRAM technique (Rosolowsky et al. 2008), implemented in the Python library `ASTRODENDRO`, constructs two-dimensional map of the hierarchical cloud structure. The largest object of the hierarchy is called *root.* Each node in the dendrogram tree splits to exactly two substructures and a sequence of nodes is called *branch.* At the top of each branch are two *leaves* (associated with intensity maxima): objects without substructures.

We compose dendrogram trees in Rosette varying two input parameters of the technique (Fig. 1). Lower intensity limit T_0 (in units K) defines the level above which extraction of the trunk is allowed and, hence, sets up the largest scale in the tree. The second input parameter ΔT_{max} represents the minimal intensity difference between the level of a node and the levels of its *both* substructures (Fig. 1). We vary T_0 and ΔT_{max} in order to select samples of dendrogram objects which: i) are rich enough; and ii) are free of objects associated with `spikes' due to noise. The chosen T_0 is different for different tracers in view of their various noise levels. Variation of ΔT_{max} controls the number of structures – decreasing the value of this parameter leads to extraction of more structures.

We chose $T_0 = 1$ for the ^{12}CO map and $T_0 = 2$ for the ^{13}CO and dust maps. The authors of the DENDROGRAM technique do not recommend values $\Delta T_{max} < 2$ in order to avoid the noise `spikes'. We opt for $\Delta T_{max} = 1$ only for the *Herschel* map – due to the low number of substructures and the need of better statistics for our method. It should be pointed out also, that the noise values in this case are low.

The samples of dendrogram objects selected for derivation of fractal dimensions are:

- ^{12}CO : $T_0 = 1$, $\Delta T_{max} = 2$ (369 objects)
- ^{13}CO : $T_0 = 2$, $\Delta T_{max} = 3$ (2075 objects)
- Dust: $T_0 = 2$, $\Delta T_{max} = 1$ (144 objects)

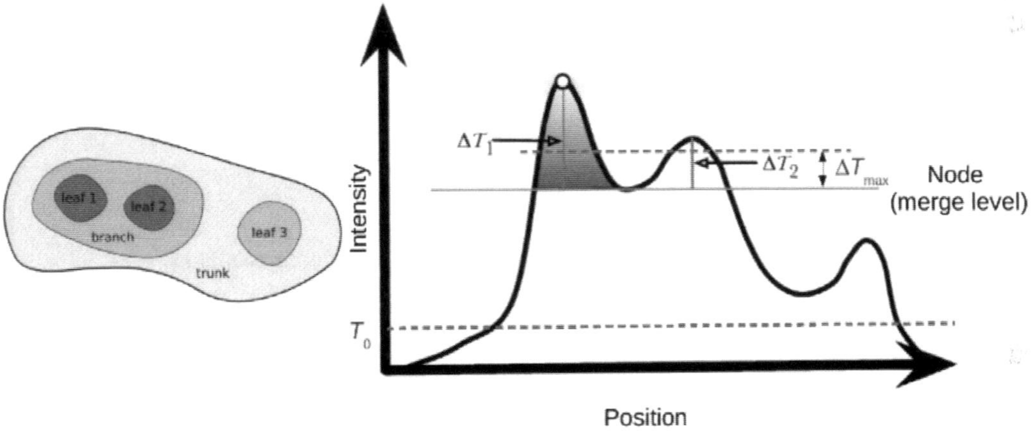

Figure 1: On the construction of the dendrogram tree from an integrated intensity map and the choice of the parameters T_0 and ΔT_{max} (see text). Only bifurcations with $\Delta T_1 > \Delta T_{max}$ and $\Delta T_2 > \Delta T_{max}$ are identified as nodes; the first such node with $T > T_0$ is identified as the root of the tree.

4. APPROACH TO DERIVE MASS DIMENSION D_M

Various approaches to calculate fractal dimensions for different structures are found in the literature. Most of them are essentially geometrical, e.g. the classical dimension of 3D fractals $D_f = \log N / \log S$, where N is the number of substructures at given scale and $S > 1$ is the scaling factor (Elmegreen, 1997;

Sánchez et al., 2005), and the box-counting dimension D_{BC}, which measures the coverage of the hierarchical object by grids of decreasing box size. Such methods account rather for geometrical properties and not for the physics behind the observed hierarchical structure properties of the objects.

More reliable approaches to describe the fractal structure of MCs are methods for derivation of the so-called *mass-length dimension* D_M, defined as the power-law exponent of the mass-size relation $M \propto L^{D}$ (Mandelbrot, 1983). Such a method was applied by Beattie et al. (2019, hereafer, BFK), using surface-density map of simulated turbulent MCs without self-gravity. The length scales in their approach are defined as increasing sizes of embedded boxes starting from a chosen peak (Fig. 2, left).

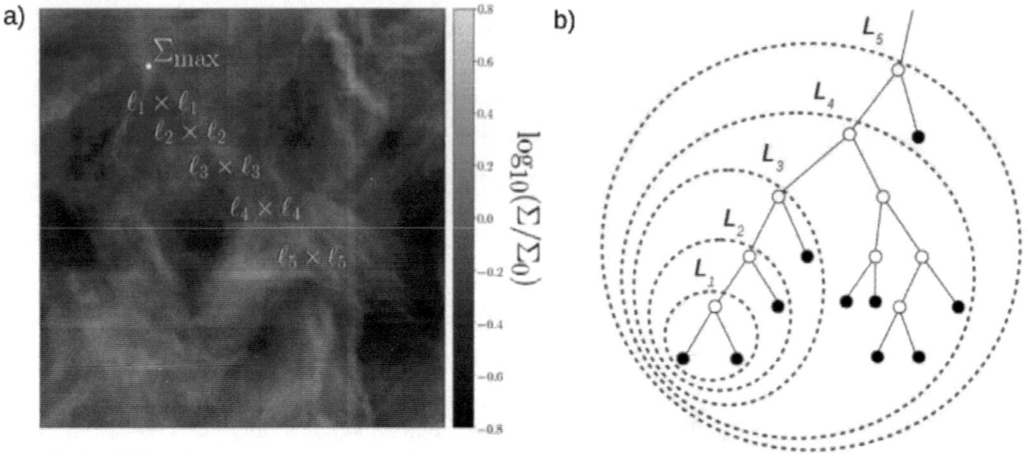

Figure 2: Comparison between two approaches to derive D_M from integrated-intensity (or column-density) maps: a) BFK (see Fig. 2 there); b) this report. See text for details.

We suggest a method to derive the fractal dimensions D_M which is similar to that of BFK but designed to be applied to a dendrogram tree. A pair of leaves (associated with some local integrated-intensity peaks) and the node they belong to are taken as the bottom of a sequence of length scales $L_1 < L_2 < \ldots < L_n$ where each L_i is the effective size of a node in the chosen branch of the dendrogram tree (Fig. 2, right) and $L_n = L_{max}$ is the effective size of the root (which depends on the choice of the input parameter T_0). The fractal dimension $D_M(L_i)$ at given scale L_i is calculated as the slope of power-law fit performed on the data set $[(L_i, M_i);(l_j, m_j)]$ where (l_j, m_j) are the sizes and masses of all substructures included in the node (L_i, M_i), i.e. of all objects within the corresponding dashed blue line in Fig. 2, right. In that way the method takes into account the contribution of the other branch that merges with the traced one in given node of size L_i.

5. SCALE DEPENDENCE OF D_M IN ROSETTE

The results of our calculations from the $^{12}CO/^{13}CO$ maps are presented in Fig. 3. The length scales are normalized to the effective size L of the root in the chosen sample of dendrogram objects – this structure includes the whole star-forming region (on the ^{12}CO map) and only the Rosette MC + the Monoceros ridge (on the ^{13}CO map). The curves $D_M=D_M(l/L)$ derived by BFK for Mach numbers 1 (transition to transonic regime) and 20 (highly supersonic regime) are plotted for comparison.

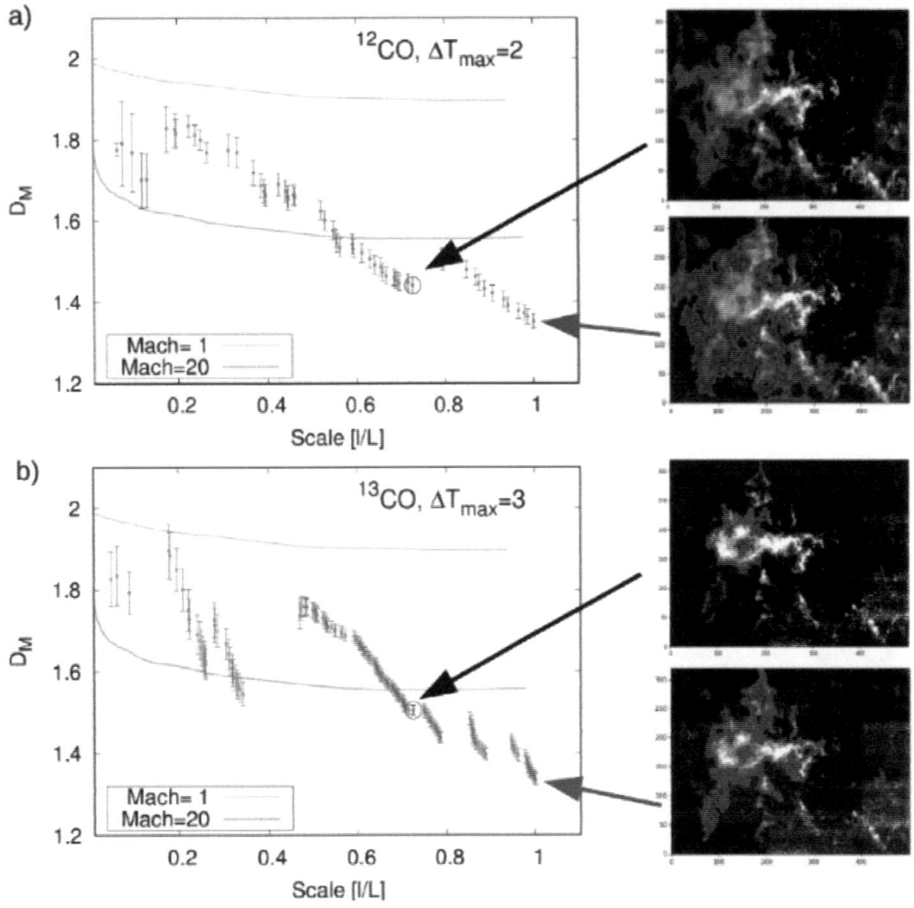

Figure 3: Scale dependence of D_M in the star-forming region Rosette as derived from: a) ^{12}CO maps; b) ^{13}CO maps. The largest scale ($l=L$) and the substructure which includes Rosette MC and the Monoceros ridge are shown with arrows.

In general, the derived scaling dependence of D_M from the $^{12}CO/^{13}CO$ maps (Fig. 3) is consistent with the numerical models of BFK in the scale range $l/L<0.5$. These dendrogram objects are mostly structures in the main ridge of the cloud which are probably not influenced by feedback effects (gas compression, heating) from the OB cluster NGC 2244 located in the bottom-right corner of the maps in

Fig. 3. At small scales (l/L<0.2) D_M~1.8, in broad agreement with the classical value D_M~2 found by Larson (1981) for a large sample of MCs and their substructures. At larger scales the derived values of D_M decrease quasi-monotonically in the ^{12}CO case and with notable discontinuities in the ^{13}CO case. The general trend hints at multifractal cloud structure. The difference between the scale dependences of D_M derived from the CO tracers can be attributed to the optical thickness of the ^{12}CO emission at high densities (Draine 2011). Thus ^{12}CO traces regions of lower column density and may fail to identify dense structures at small scales. Probably this is the reason for the smoother scale dependence of the fractal dimension from the ^{12}CO sample compared to ^{13}CO (Fig. 3).

There are at least two possible explanations of the inconsistency of our results at large scales (l/L~0.7 and larger) with the numerical study of BFK. First, the largest structures include zones outside the Rosette MC – mostly from the Monoceros Ridge – where the physical conditions differ substantially. Second, the number of dendrogram objects per bin in this scale range obviously increases which affects the output of the fitting procedure. Probably this is an artifact of the DENDROGRAM technique itself and needs to be clarified in a forthcoming study.

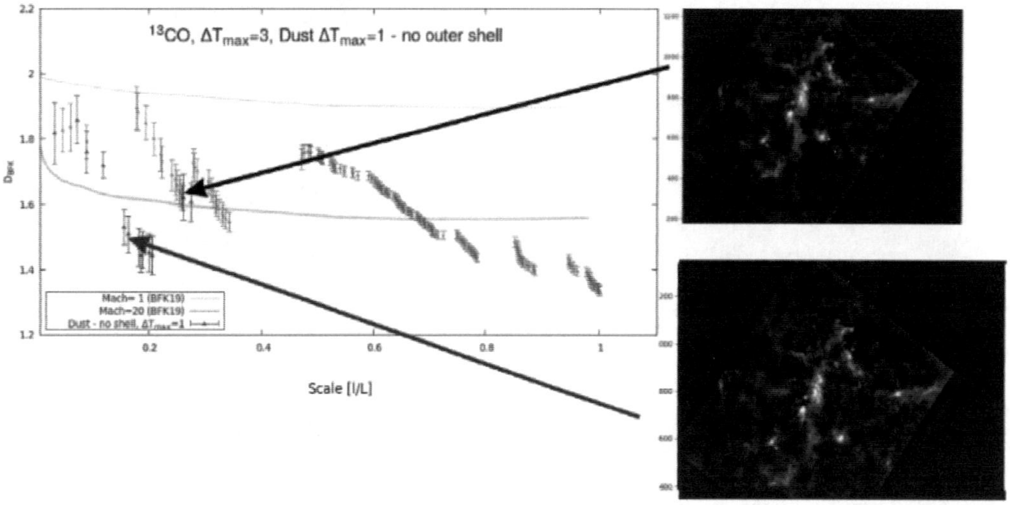

Figure 4: Scale dependence of D_M in Rosette MC derived from *Hershel* maps compared with the results from ^{13}CO maps. Structures in the main ridge of the cloud are shown with arrows.

In Fig. 4 we compare the scale dependence of D_M derived from the *Hershel* maps with the results from the ^{13}CO tracer. Dust emission is optically thin which allows investigation of denser fragments of the Rosette MC as seen on the map (Fig. 4, right). Unfortunately, the sample size is small and this affects significantly the derivated values of D_M. The smallest dust structures yield fractal dimensions in agreement with the results from the ^{12}CO/^{13}CO samples while the discontinuity of the scale dependence of D_M is probably an effect of the sample size.

6. CONCLUSION

We suggest a method to estimate the fractal dimensions D in molecular clouds by use of the mass-size relation $M(L)\sim L^{-D}$ where the power index $D=D_M$ is called mass-length dimension. In contrast to the similar approach of Beattie et al. (2019), the technique is not based on abstract scales but deals with samples of dendrogram objects with their effective sizes.

The star-forming region Rosette was chosen as an object to test the method, by use of maps in molecular-line ($^{12}CO/^{13}CO$) and dust-emission tracers. The derived scale dependence of D_M in Rosette MC is in general agreement with the numerical study of Beattie et al. (2019) for molecular clouds in transonic up to highly supersonic regime. The inconsistency found at larger scales which correspond to the whole star-forming region could be explained by changes in the physical conditions and/or with artifacts of the DENDROGRAM technique. This needs to be clarified in a forthcoming study.

Acknowledgements

The authors are grateful to M. Heyer and J. Williams for the FCRAO molecular-line maps and to N. Schneider for the *Herschel* maps of Rosette given at our disposal and used in this report.

M. Bogdanova thanks to the Bulgarian National Science Fund for providing support through Grant KP-06-PM-38/6 (Fundamental research by young scientists and postdocs 2019). T. Veltchev acknowledges funding from the Ministry of Education and Science of the Republic of Bulgaria, National RI Roadmap Project DO1-277/16.12.2019.

References

Andre P. et al.: 2010, *Astronomy & Astrophysics*, 518, L102.

André P., Di Francesco J., Ward-Thompson D., Inutsuka S.-I., Pudritz R., Pineda J.: 2014, in Beuther H., Klessen R. S., Dullemond C., Henning T., eds, *Protostars and Planets VI*. Univ. Arizona Press, Tucson, p. 27.

Beattie J., Federrath C., Klessen R. S.: 2019, *Monthly Notices of the Royal Astronomical Society*, 487, 2070 (BFK).

Bergin E. A., Tafalla M.: 2007, *Annual Review of Astronomy and Astrophysics* , 45, 339.

Dent W. et al.: 2009, *Monthly Notices of the Royal Astronomical Society*, 395, 1805.

Di Francesco J. et al.: 2010, *Astronomy & Astrophysics*, 518, L91.

Draine B.& 2011, *Physics of the Interstellar and Intergalactic Medium*. Princeton Univ. Press, Princeton, NJ.

Elmegreen B. G.: 1997, *The Astrophysical Journal*, 486, 944.

Elmegreen B. G., Falgarone E.: 1996, *The Astrophysical Journal*, 471, 816.

Heyer M., Brunt C.: 2004, *The Astrophysical Journal*, 615, L45.

Heyer M., Krawczyk C., Duval J., Jackson J.: 2009, *The Astrophysical Journal*, 699, 1092.

Heyer M., Williams J., Brunt C.: 2006, *The Astrophysical Journal*, 643, 956.

Larson R.: 1981, *Monthly Notices of the Royal Astronomical Society*, 194, 809.

Lombardi M., Alves J., Lada C.: 2011, *Astronomy & Astrophysics*, 535, A16.
Mandelbrot B. (1983). The Fractal Geometry of Nature. W. H. Freeman and Co., New York, United States of America. ISBN 0-7167-1186-9.
Rosolowsky E., Pineda J., Kauffmann J., Goodman A.: 2008, *The Astrophysical Journal*, 679, 1338.
Sánchez N., Alfaro E. J., & Pérez E.: 2005, *The Astrophysical Journal*, 625, 849.
Schneider N., Stutzki J., Winnewisser G., Block D.: 1998, *Astronomy & Astrophysics*, 335, 1049.
Schneider N., Motte F., Bontemps S., Hennemann M., et al.: 2010, *Astronomy & Astrophysics*, 518, L83.
Schneider N., Csengeri T., Hennemann M., Motte F., et al.: 2012, *Astronomy & Astrophysics*, 540, L11.
Veltchev T., Ossenkopf-Okada V., Stanchev O., Schneider N., Donkov S., Klessen R. S.: 2018, *Monthly Notices of the Royal Astronomical Society*, 475, 2215.
Williams J., Blitz L., Stark A.: 1995, *Astrophysical Journal*, 451, 252.

Proceedings of the XII Serbian-Bulgarian Astronomical Conference (XII SBAC) Sokobanja, Serbia, September 25-29, 2020
Editors: L. Č. Popović, V. A. Srećković, M. S. Dimitrijević and A. Kovačević
Publ. Astron. Soc. "Rudjer Bošković" No 20, 2020, 69-77

TRACING THE LOCAL MORPHOLOGY OF THE MOLECULAR CLOUD ROSETTE USING MOLECULAR-LINE DATA

ORLIN STANCHEV, TODOR V. VELTCHEV and MARIYANA BOGDANOVA

Department of Astronomy, University of Sofia, James Bourchier Blvd. 5, Sofia 1164, Bulgaria
E-mail: mpetrova@phys.uni-sofia.bg, o_stanchev@phys.uni-sofia.bg, eirene@phys.uni-sofia.bg

Abstract. Recent kinematic and high-resolution observational studies of molecular clouds reveal their extremely complex, clumpy and filamentary structure. Maps of different molecular-line tracers and dust-opacity data, combined with appropriate clump-extraction methods, allow for investigation of the local morphology and its physical interpretation, e.g. to test the gravoturbulent scenario of cloud evolution. We perform a comparative analysis of two populations of clumps in the molecular cloud Rosette extracted from ^{12}CO and ^{13}CO-emission FCRAO maps. The used extraction methods differ in their concept: whether the clumps are considered as an ensemble of independent objects (GAUSSCLUMPS) or as a hierarchical set of embedded structures (DENDROGRAM). We derive basic physical characteristics of the clumps and analyze their scaling relations.

1. INTRODUCTION

Giant molecular clouds (MCs) with masses up to $10^{4\text{-}5}$ solar masses are the places where most of the stars are born in clusters shrouded by cold molecular gas and dust. They contain many smaller and denser fragments or clumps whose properties are determined by the interplay between gravity, magnetic fields and turbulent pressure. Stars form in such clumps; thus the derivation and analysis of their physical properties are important tasks for the better understanding of star formation processes.

A dominant characteristics of MCs is their hierarchical structure. Many multitracer studies have shown that the relatively small scales in that hierarchy correspond to high-density features and are invariably contained inside envelopes of lower-density gas (e.g., Blitz and Shark 1986, Lada 1992). Any given spatial scale contains more small dense structures than large sparse structures. Dense cores with size ~0.1 pc constitute the top level of the cloud hierarchy and correspond to the scale of transition to coherent turbulence (Tafalla et al. 2004; Lada et al. 2008). Despite of the non-uniformity of the filling factor and the chemical state of MCs, the bottom of gas hierarchy can be attributed to low-density gas that fills most of the cloud volume. However, there are two main interpretations connecting the boundaries of MCs with the bottom of hierarchy. The "classical" one (Blitz et al. 2007) suggests that the cloud boundaries form clearly distinguishable bottom level of the structure hierarchy. The alternative interpretation is that the hierarchical structure continues only with chemical changes into the low-density diffuse ISM (Ballesteros-Paredes et al. 1999; Hartmann et al. 2001). Crucial factor in favor of some of these interpretations could be the lifetimes of MCs in relation to their internal crossing times, pointing to the importance of self-gravity in cloud energetics (Elmegreen 2007).

A broadly accepted approach in analysis of molecular-line data is the segmentation of the position-position-velocity (PPV) cube into physically relevant structures (population) and derivation of their characteristics. For instance, such cloud fragments/clumps are identified as connected regions of emission above a threshold intensity (Solomon et al. 1987). Some of the mostly used segmentation methods are the clump-extraction algorithms CLUMPFIND (Williams et al. 1994; water-shed segmentation technique), GAUSSCLUMPS (Stutzki & Güsten 1990; iterative fitting of three-dimensional Gaussians in the vicinity of peaks on the residuum map) and DENDROGRAM (Rosolowsky et al. 2008; application of structure trees to reveal the hierarchical connections between different size scales in MCs).

In this report we present a continuation of our analysis of clump populations in the Rosette MC derived by use of GAUSSCLUMPS from $^{12}CO/^{13}CO$ and *Herschel* data (Veltchev et al. 2018). Here we derive physical properties of the clump population extracted by use of the DENDROGRAM technique and compare them with those of the population of Gaussian clumps

2. OBSERVATIONAL DATA

The star-forming cloud Rosette is an appropriate object for such comparative study. Its local morphology has been intensively investigated by use of different clump-extraction methods and tracers of the molecular gas (Williams et al. 1995; Schneider et al. 1998; Dent et al. 2009; Di Francesco et al. 2010; Veltchev et al. 2018). In this work, we do not consider the expanding HII region around the young cluster NGC 2244 and the zone of its direct interaction between it and the Rosette MCs which is called 'Monoceros Ridge' (Blitz & Thaddeus 2010). Various effects of stellar feedback such as gas compression by the expanding ionization front or radiation heating from the cluster can influence physical properties of the clump population in the Monoceros Ridge region. While some studies (Schneider et al. 2012; Cambresy et al. 2013) showed that there is no indication for large-scale triggering of star-formation further inside Rosette cloud, other reveals that more massive dense cores forms in this zone (Motte et al. 2010). The adopted distance to Rosette MC is 1.33 kpc (Lombardi, Alves & Lada 2011). We make use of $^{12}CO/^{13}CO$ PPV cubes taken with the 14 m telescope of Five College Radio Astronomy Observatory (FCRAO), presented and discussed by Heyer, Williams & Brunt (2006). The spectral resolutions are 0.127 km/s (^{12}CO data) and 0.133 km/s (^{13}CO data) and the angular resolution is 46 arcsec. All temperatures are given on the main beam brightness temperature scale.

3. CLUMP EXTRACTION USING DENDROGRAM METHOD

To extract the population of Dendrogram clumps, we use the PYTHON implementation of the DENDROGRAM technique called ASTRODENDRO. A dendrogram structure can be graphically represented as a tree of hierarchical objects in the data cube (Fig. 1). The tree is composed of two types of structures: *branches* or *nodes* (objects consisting of two substructures) and *leaves* (objects without substructures, associated with intensity maxima). The largest structure in the hierarchical tree is the *trunk* (or *root*).

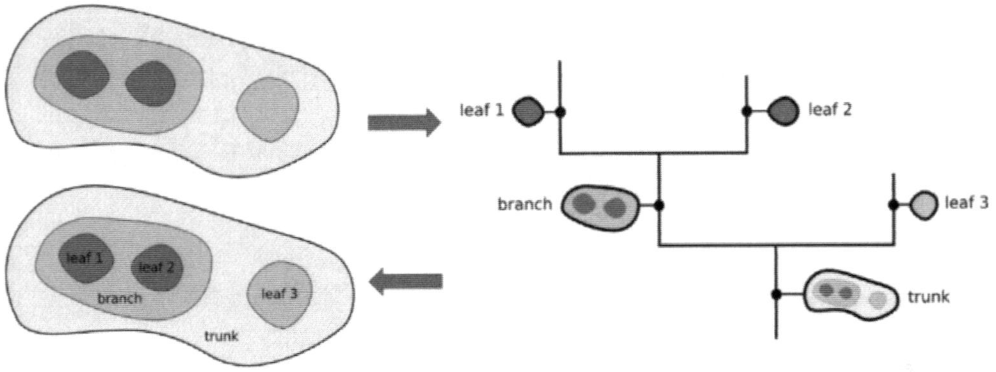

Figure 1: Illustration of a dendrogram: an arbitrary hierarchical structure (left) and the corresponding dendrogram tree (right). The trunk is the largest structure in the hierarchy and contains all other structures.

ASTRODENDRO identifies unique isosurfaces from each region in the PPV data cube and computes the properties of the delineated clump using the moment of volume weighted intensities of emission from every pixel (Rosolowsky et al. 2008). The rms sizes along the major and minor clump axis are computed from the intensity-weighted second moment in two dimensions, *dx* and *dy,* and then the effective radius is defined as the geometric mean of these second spatial moments. In a similar way, the velocity dispersion is the intensity-weighted second moment of the velocity axis. The flux of a clump is the sum of all emission within an isosurface:

$$F = \sum T_i dx dy dv,$$

where T_i is the brightness temperature. Thus, clump luminosity can be calculated as the integrated flux scaled by the square of the distance to the object:

$$L_{CO}[\mathrm{K\ km\ s^{-1}\ pc^2}] = D^2 \sum (T) dx dy dv$$

The ASTRODENDRO implementation of the DENDROGRAM technique uses three input parameters that determines the hierarchy of extracted objects: the minimum value to be considered in the data set (*min_value*), the threshold value that determines whether a leaf is a single entity or not (*min_delta*), and the minimal number of pixels for a leaf to be considered as a single entry (*min_npix*). By varying *min_value* and *min_delta,* we generated multiple clump samples to study the method's sensitivity to the choice of input parameters. As expected, large values of *min_value* produce small samples of clumps filling a reduced volume of the data cube. On the other hand, choosing too low *min_value* may lead to identification of some noise spikes in the data as physical dendrogram objects; therefore *min_value* < 2 is not recommended by the authors of the DENDROGRAM technique. The choice of *min_delta* controls how significant a leaf has to be in order to be considered as an independent entity. The measure of significance is the difference between its peak flux and the value at which it is being merged into the structure tree. Thus, the choice of *min_delta* affects the number of extracted leaves in the extracted clump sample. The default value of *min_npix* is set to six. A leaf which consists of less pixel as the dendrogram tree is being constructed is merged with a branch or another leaf and is not considered as a separate entity.

4. SPATIAL ASSOCIATION BETWEEN GAUSSIAN AND DENDROGRAM CLUMP POPULATIONS

Various tracers of cloud structure are sensitive to different ranges of density and optical depth while the specifics of a chosen clump-extraction algorithm reflect the physical processes behind the observed local cloud morphology. Therefore, spatial association (or, cross-identification) between clumps ensembles derived from different tracers and/or obtained from different clump-extraction techniques might be helpful to understand the physics of MCs. We follow the technique described in Veltchev et al. (2018) to associate Gaussian and dendrogram clump populations in the Rosette MC. Fig. 2 displays the spatial distribution of dendrogram clumps (light grey) for the choice of input parameters (*min_value* = 2 sigma, *min_delta* = 2 sigma) juxtaposed with that of Gaussian clumps (dark grey) extracted from the ^{12}CO (left) and ^{13}CO map (right). The region which includes the Monoceros ridge and the cluster NGC 2244 (shaded circle) is excluded from further analysis (see Sect. 2). As seen in Fig. 2, both Gaussian and dendrogram populations trace the `main ridge' of the Rosette MC and some of the larger filaments.

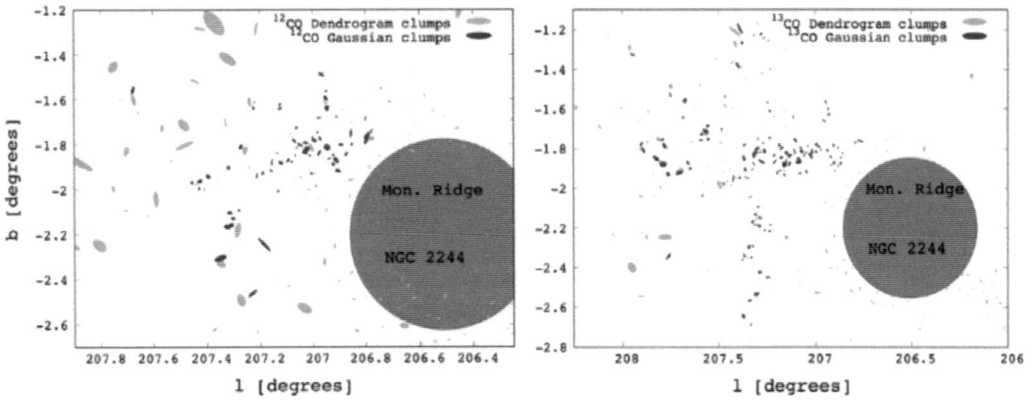

Figure 2: Spatial distributions of Gaussian (black color) and Dendrogram (grey color) clump populations for 12CO and 13CO PPV data cubes. The Monoceros Ridge zone is labeled in circle (bottom right side on the diagrams).

Let us briefly recall the technique Veltchev et al. (2018) for cross-identification between the clump populations. The overlap coefficient w between 2 clumps is defined as the intersection area, normalized to the area of the smaller ellipse (assuming elliptical geometry of the clumps). Clump pairs with $50\ \% \leq w < 90\ \%$ are defined as overlapping (type O) while those with $w \geq 90\ \%$ – as embedded (Type E). Such criteria are strong enough to ensure that most of the mass of a centrally condensed clump is contained in the overlap area. The subsequent physical analysis showed the ability of this approach to reveal essential properties of the cloud structure. We applied it for cross-identification between Gaussian and dendrogram clumps and summarize the results in Table 1.

	Total number of Gaussian clumps	Total number of Dendrogram clumps	Associated Gaussian clumps	Associated Dendrogram clumps	Number ratio (E / O)
^{12}CO	68	102	18	17	0.8
^{13}CO	130	163	51	44	1.32

In a follow-up paper we intend to provide a more detailed statistics on the cross-identification between the Gaussian and Dendrogram clump populations.

5. SCALING RELATIONS

Physical parameters of clumps in Galactic MCs and their relation to cloud structure and star formation have been investigated in numerous works. Many of these studies are dedicate on scaling relations between observable quantities like velocity dispersion and mass. In his seminal analysis, Larson (1981) derives three empirical properties of clouds and their substructures (clumps) in the Milky Way: 1) a power-law relation between velocity dispersion and size of emitting medium: $\sigma_v \propto R^{0.38}$; 2) virialization of MCs: $2\sigma_v R^2/GM \sim 1$; 3) a power-law scaling relation of the mean cloud/clump density: $n \propto R^{-1.1}$. These relations have been later put to test through observations in different molecular tracers including rotational transitions of CO and its isotopes as well tracers of dense gas like NH_3, CS, and HCN (Dame et al. 1986; Heyer et al. 2001, 2009). Many of those studies reveal power-law scaling relation of the velocity dispersion with index ranging from 0.25 to 0.75 indicating self-gravitating cloud fragments in virial equilibrium.

Fig. 3 shows the size vs. mass relations for dendrogram leaves (empty circles) from the ^{12}CO and ^{13}CO map, plotted for comparison with the Gaussian clumps (filled cirlcles) analyzed by Veltchev et al. (2018). The dendrogram samples were obtained setting the input parameters *min_value* = *min_delta* = 2σ. The sizes of the dendrogram objects were calculated adopting definition of cloud radius by Solomon et al. (1987):

$$R \approx 1.91\,\sigma_r,$$

where σ_r is the geometric mean of the second spatial moments along the major and minor axis of the clump. Then the virial mass of the dendrogram clumps is calculated as follows:

$$M_{\rm vir} = \frac{5\,FWHM_v\,R}{2G},$$

where the $FWHM_\upsilon$ is calculated as:

$$FWHM_v = 2\sqrt{2\ln 2}\sqrt{\sigma_v^2 + \left(\frac{\Delta v}{2\sqrt{2\ln 2}}\right)^2}$$

where the second term under the square root represents the velocity resolution term with $\Delta\upsilon$ being the velocity channel width (1.127 km/s for ^{12}CO data and 0.13 km/s for ^{13}CO data).

The masses of dendrogram objects in ^{12}CO emission seem to be systematically higher than those of the Gaussian clumps. This tendency is more pronounced for the population of leaves. A possible explanation could be the different nature of the dendrogram and Gaussian objects. While a 3D Gaussian is a more appropriate representation of high-density clumps which are better traced by the ^{13}CO emission, the dendrogram objects represent cloud fragments in their real complex shapes. The same holds even more for ^{12}CO data which are sensitive to lower density regimes in MCs and the extracted clumps from this tracer cover more diffuse regions that possibly can contain much more mass.

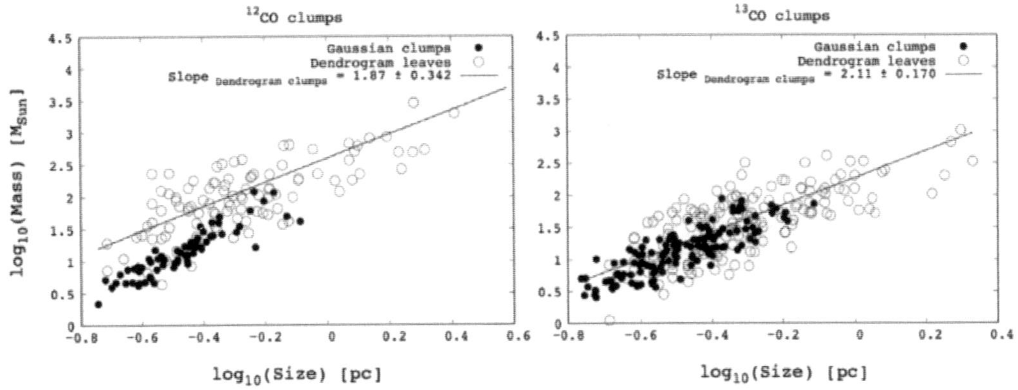

Figure 3: Size vs. mass relations for the dendrogram leaves (empty circles) and the Gaussian clumps (filled circles) from Veltchev et al. (2018). The best-fit lines for the dendrogram population are shown.

Fig. 4 displays the corresponding size vs. linewidth relations for dendrogram leaves and the associated Gaussian clumps. In general, the correlation is weak but the power index increases for the sample of dendrogram leaves associated with the Gaussian clumps from given tracer.

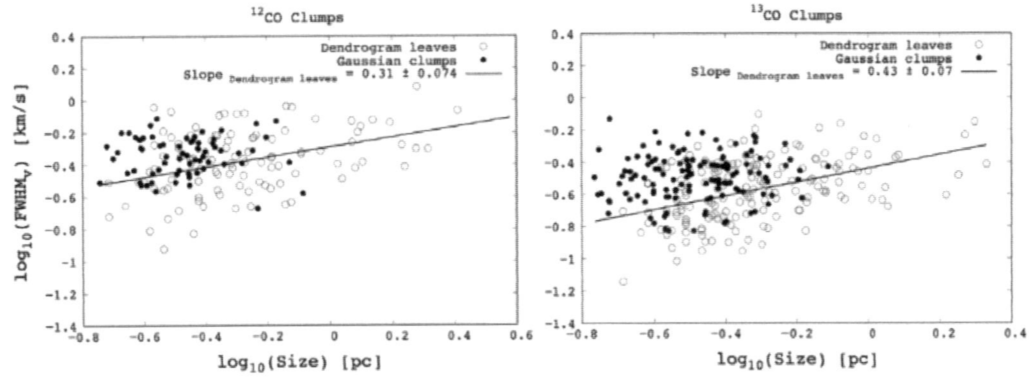

Figure 4: Size vs. linewidth relations for the dendrogram leaves (empty circles) and the Gaussian clumps (filled circles) from Veltchev et al. (2018). The best-fit lines for the dendrogram population are shown.

The summarized results for the determined slopes of the size vs. mass and size vs. linewidth relations and for three different clump populations (all dendrogram objects, dendrogram leaves only and dendrogram leaves associated with Gaussian counterparts) is shown in Figure 5.

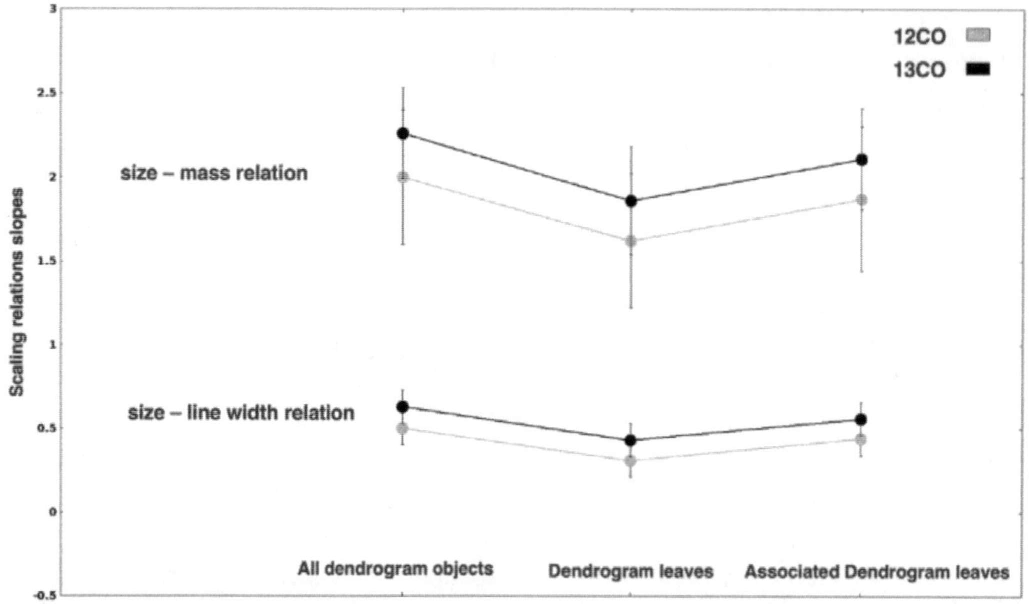

Figure 5: Summary on the derived indexes of scaling relations for clump populations extracted from the ^{12}CO and ^{13}CO maps.

6. CONCLUSION

We performed a comparative analysis of the derived properties of clump populations extracted from maps of ^{12}CO and ^{13}CO emission in the Rosette molecular cloud. Adopting the alternative extraction techniques GAUSSCLUMPS (clumps are considered as an ensemble of independent objects) and DENDROGRAM (clumps are considered as a hierarchy of embedded structures), we study the local cloud morphology, derive some basic physical characteristics of clumps and analyze their scaling relations.

The dendrogram populations exhibit shallower scaling of mass in comparison with the Gaussian clump populations whose scaling index exceeds 2.0 ± 0.3. The size vs. mass relation for the ^{12}CO populations differs from that of the ^{13}CO populations. The masses of the dendrogram objects calculated from both tracers are systematically higher than those of the Gaussian clumps; this tendency is more pronounced when one considers only the dendrogram leaves. The weak correlation of velocity dispersion with size for the dendrogram objects could be indicative of their evolutionary state.

The associated Gaussian clumps tend to have relatively high linewidths, especially considering the ^{12}CO population.

This study could be extended including the Monoceros Ridge region where the stellar feedback effects must be taken into account.

Acknowledgement

The authors are grateful to M. Heyer and J. Williams for the FCRAO molecular-line maps. This research made use of ASTRODENDRO, a Python package to compute dendrograms of Astronomical data (http://www.dendrograms.org). This research also made use of ASTROPY (http://www.astropy.org). O. Stanchev thanks to the Bulgarian National Science Fund for providing support through Grant KP-06-PM-38/6 (Fundamental research by young scientists and postdocs 2019). T. Veltchev acknowledges funding from the Ministry of Education and Science of the Republic of Bulgaria, National RI Roadmap Project DO1-277/16.12.2019.

References

Ballesteros-Paredes J., Va´zquez-Semadeni E., Scalo, J.: 1999, *ApJ*, 515, 286.

Blitz L., Fukui Y., Kawamura A., Leroy A., Mizuno N., Rosolowsky E.: 2007, in Protostars and Planets V, ed. B. Reipurth, D. Jewitt and K. Keil, Tucson, Univ. Arizona Press, 81.

Blitz L., Thaddeus P.: 1980, *ApJ*, 241, 676.

Blitz L., Stark A. A.: 1986, *ApJ*, 300, L89.

Cambresy L., Marton G., Feher O., T´oth L., Schneider N.: 2013, *A&A*, 557, 29.

Dame T. M., Elmegreen B. G., Cohen R. S., Thaddeus P.: 1986, *ApJ*, 305, 892.

Dent W. et al.: 2009, *Monthly Notices of the Royal Astronomical Society*, 395, 1805.

Di Francesco J. et al.: 2010, *Astronomy & Astrophysics*, 518, L91.

Elmegreen B. G.: 2007, *ApJ*, 668, 1064.

Hartmann L., Ballesteros-Paredes J., Bergin E. A.: 2001, *ApJ*, 562, 852.

Heyer M. H., Carpenter J. M., Snell R. L.: 2001, *ApJ*, 551, 852.

Heyer M., Krawczyk C., Duval J., Jackson J. M.: 2009, *ApJ*, 699, 1092, arXiv: 0809.1397.

Heyer M., Williams J., Brunt C.: 2006, *The Astrophysical Journal*, 643, 956.

Houlahan P., Scalo J.: 1990, *ApJS*, 72, 133.

Lada E. A.: 1992, *ApJ*, 393, L25.

Lada C. J., Muench A. A., Rathborne J. M., Alves J. F., Lombardi M.: 2008, *ApJ*, 672,.410.

Motte F., Zavagno A., Bontemps S., Schneider N., Hennemann M., Di Francesco J., Andre Ph., Saraceno P., et al.: 2010, A&A, 518, L77.

Rosolowsky E., Pineda J., Kauffmann J., Goodman A.: 2008, *The Astrophysical Journal*, 679, 1338.

Schneider N., Stutzki J., Winnewisser G., Block D.: 1998, *Astronomy & Astrophysics*, 335, 1049.

Schneider N., Csengeri T., Hennemann M., Motte F., et al.: 2012, *Astronomy & Astrophysics*, 540, L11.

Stutzki J., Güsten R.: 1990, *ApJ*, 356, 513.

Tafalla M., Myers P. C., Caselli P., Walmsley C. M.: 2004, A&A, 416, 191.

Veltchev T., Ossenkopf-Okada V., Stanchev O., Schneider N., Donkov S., Klessen R. S.: 2018, *Monthly Notices of the Royal Astronomical Society*, 475, 2215.

Williams J. P., de Geus E. J., Blitz L.: 1994, *ApJ*, 428, 693.

Williams J., Blitz L., Stark A.: 1995, *Astrophysical Journal*, 451, 252.

Proceedings of the XII Serbian-Bulgarian Astronomical Conference
(XII SBAC) Sokobanja, Serbia, September 25-29, 2020
Editors: L. Č. Popović, V. A. Srećković, M. S. Dimitrijević and A. Kovačević
Publ. Astron. Soc. "Rudjer Bošković" No 20, 2020, 79-84

SHADOW BANDS AND RELATED ATMOSPHERIC PHENOMENA REGISTERED DURING TOTAL SOLAR ECLIPSES

RUSLAN ZLATEV[1], NIKOLA PETROV[1], TSVETAN TSVETKOV[1], EMIL IVANOV[2], ROSITSA MITEVA[1], VELIMIR POPOV[2], YOANA NAKEVA[3] and LJUBE BOJEVSKI[4]

[1]*Institute of Astronomy and National Astronomical Observatory, Bulgarian Academy of Sciences, 72 Tsarigradsko Shose blvd., Sofia 1784, Bulgaria*
[2]*Konstantin Preslavsky University of Shumen, 115 Universitetska str., Shumen 9700, Bulgaria*
[3]*Aix-Marseille University, 3 Place Victor Hugo, 13003 Marseille, France*
[4]*Ss Cyril and Methodius University, 9 Goce Delcev blvd., Skopje 1000, Republic of North Macedonia*
E-mail: tstsvetkov@astro.bas.bg

Abstract. This study is dedicated to the atmospheric phenomena accompanying total solar eclipses. Observations of shadow bands are shown. We look for a connection between their distribution and the variations of the temperature, speed and directions of the wind before, during and after the totality of two total solar eclipses. A new experiment for registering the shadow bands realized during the last total solar eclipse (2019 July 2) is presented.

1. INTRODUCTION

Total solar eclipses (TSEs) have impressed observers since the beginning of recorded history. Despite the existence of evidence for solar eclipse observations from more than 4 millennia ago, some accompanying phenomenon remained unnoticed for thousands of ages. It wasn't until 1820 that Hermann Goldschmidt noted shadow bands visible just before and after totality at some eclipses (Guillermier and Koutchmy, 1999). Later, in 1842, when George Airy saw his first TSE, he also highlighted the shadow bands (Littmann et al., 1999), but the first hypotheses that successfully explain their formation date back from the late XX century. Previously, XIXth-century observers assumed that shadow bands were some sort of diffraction phenomenon because their linear patterns roughly resemble optical interference fringes. More exotic explanations have been proposed in 1924 (Hastings, 1924). They suggest that the bands were overlapping pinhole images of the Sun formed by vents in the upper atmosphere. More recently A. L. Stanford has proposed a Lloyd's mirror effect, where direct rays from the uneclipsed solar crescent interfere with those reflected from clouds (Stanford, 1973).

But by far the simplest and most satisfactory explanation was the one published in the late 1980's by Codona (1986). It states that the shadow bands are diffraction effects caused by turbulence in the Earth's atmosphere when the light from narrow slit-like source passes through (Fig. 1). The roughly linear patterns moving across the ground with typical speeds of a few meters per second usually appear just before and just after the total phase of solar eclipses. Typically, they align parallel to the tangent to the center of the solar crescent (Marschall et al., 1984). They are easily observable by naked eye, but their low contrast makes their quantitatively measurements harder. Still, it is considered that the shadow band spacing decreases and their contrast increases as the totality approaches (Codona, 1986). Probably the first attempt for capturing shadow bands was made during a TSE in 1912 although it was rather unsuccessful. Even nowadays despite the development of technology, registration of the bands by photo or video detectors remains an uneasy task.

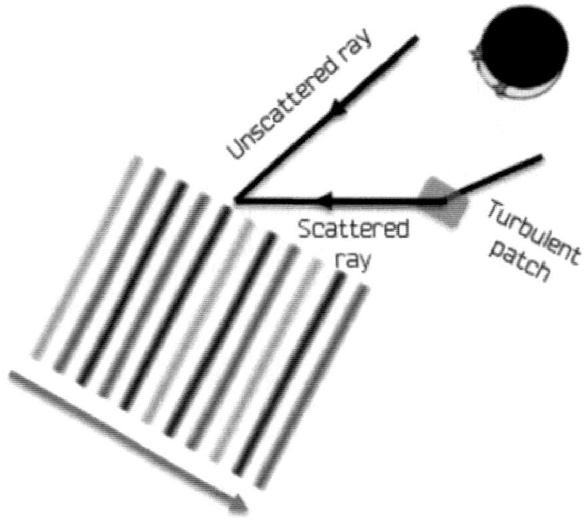

Figure 1: Formation and propagation of shadow bands.

Our team is experienced in shadow bands observations with 4 successful experiments conducted at 4 TSE expeditions – in 1999, 2006, 2017 and 2019. As the techniques and instruments used to register the bands have changed over the years, our latest results are incomparable with the earliest ones. Therefore, the current study summarizes only the data obtained during the last two TSEs (2017 August 21 and 2019 July 2).

2. EXPERIMENTAL SETUP

Our team observed the 2017 August 21 TSE from the northwestern part of the USA – the state of Oregon. We chose an observational spot (44°42'11.6"N 120°47'52.2"W) at distant, uninhabited place at altitude about 1200 m. The setup for the shadow bands experiment included white screen for better visualization and Sony DCR-SR55 digital video camera that takes 25 frames/second. Next to it at altitude 1.5 m above the ground was situated a Gill Windsonic anemometer that records the wind speed and direction 4 times/second.

We repeated the experimental setup two years later, when observing the TSE on 2019 July 2 from Atacama Desert, Chile. The location of our team (29°47'37.0"S 70°53'12.4"W) was similar as the place was far from any settlements and other observers at more than 1400 m above the sea level. The installation for registering the shadow bands and eclipse meteorology was almost the same except for the video camera that was replaced by Sony Alpha 7III that offers better quality and resolution and makes capturing the bands easier taking 50 framse/second.

3. RESULTS

During the two latest expeditions for TSE observations our team performed an experiment dedicated to registration of shadow bands and wind properties. Its aim was to look for a possible relationship between the direction of propagation and the speed of near ground wind and the observed pattern of the shadow bands. The results are summarized in Table 1. Monitoring and recording the wind characteristics and shadow bands was performed both before and after the totality of every eclipse. Unfortunately, the insufficient quality of the videotaping after the totality in 2017 does not allow data processing, thus such information is not included. The average wind speed is given only for the moments of visibility of the shadow bands – periods of about 30 seconds before and after the totality. The frequency is measured after video processing that aims to find and remove all static regions of each frame, so that small changes in the intensity became visible. This has been achieved by clearing small random noises, applying adaptive background learning and subtraction over the sequence of the frames, and amplifying the signal by increasing the contrast and brightness. A mean intensity in a fixed square region is calculated for each frame in the video and the obtained values are saved and subjected to Fast Fourier Transform.

Table 1: Properties of the wind and shadow bands registered at two TSEs – in 2017 and 2019.

	2017	2019	
	before totality	before totality	after totality
Average wind speed [m s^{-1}]	1.01	2.27	1.59
Shadow bands frequency [Hz]	1.35	1.75	1.37

It is obvious that the near ground winds during 2019 TSE were stronger than these two years earlier, but it is also noticeable that the wind speed rapidly changes before and after the totality in 2019. The observed decrement of the wind speed coincides with the decrement of the frequency of the shadow bands. The relationship between the average near ground wind speed and shadow bands frequency is shown on Fig. 2.

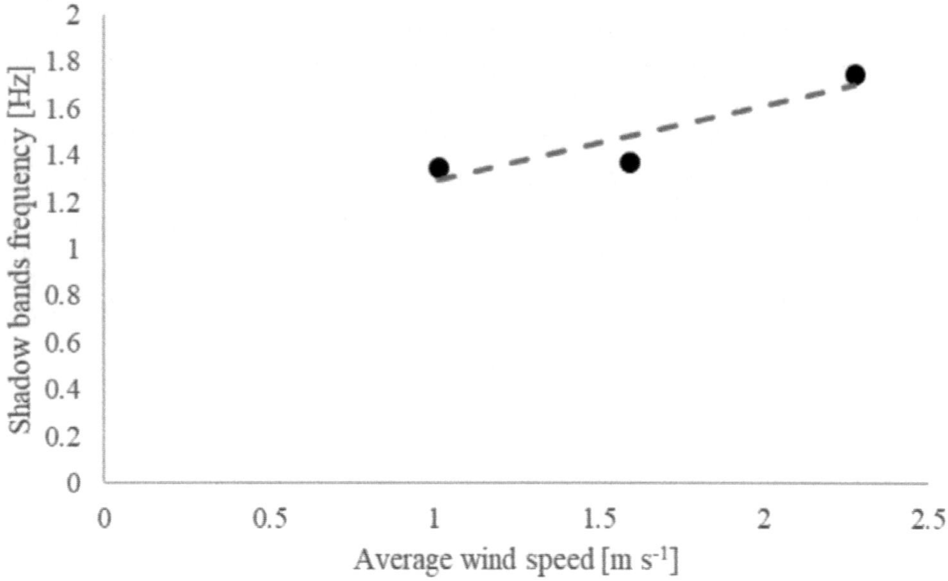

Figure 2: The relation between the average speed of the wind at the moment of shadow bands observations and their frequency.

More detailed information about the wind speed and direction during the observations of shadow bands (not only the average value) gives the wind rose diagram (Fig. 3). It shows the near ground wind properties detected only during the experiment held in Chile in 2019 and shows that the main direction of propagation is the same before and after the totality (eastern-northeastern) as the eastern component becomes weaker after it.

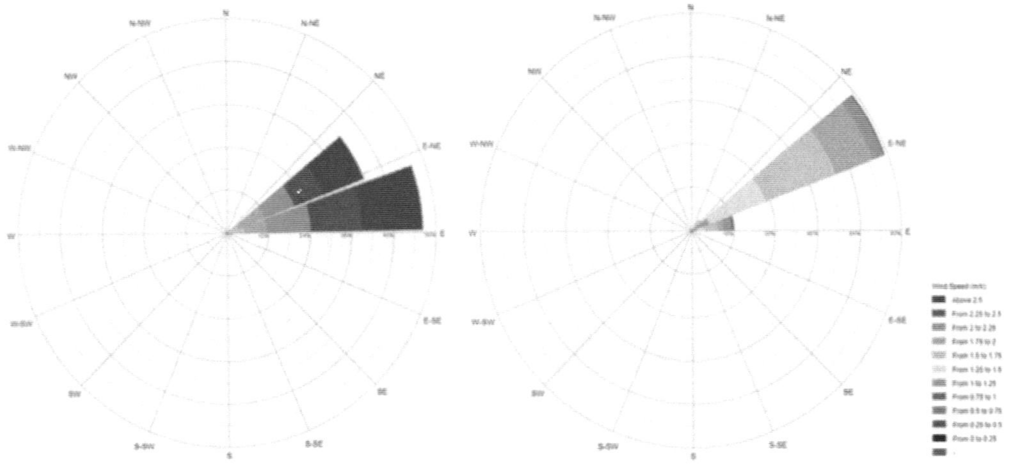

Figure 3: Wind rose diagram showing the direction of propagation and the speed of the wind detected in the periods of visibility of the shadow bands during 2019 TSE before (left panel) and after (right panel) the totality.

Two still frames from the performed videotaping of the shadow bands during the 2019 TSE (before and after the totality) are shown on Fig. 4. The background content is removed and the contrast and brightness of the pixels, which represents moving bands, are increased for better visualization. A comparison between Fig. 3 and Fig. 4 shows that the slight changes in the primary direction of propagation of the wind coincides with similar behavior of the direction of the bands.

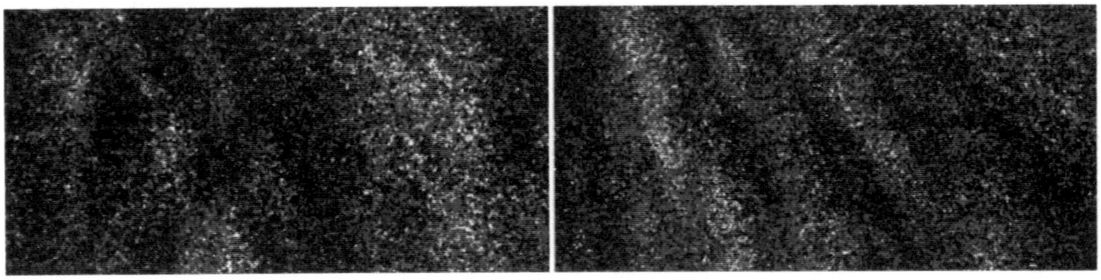

Figure 4: Two frames from the video of the shadow bands during the 2019 TSE before (left panel) and after (right panel) the totality.

4. CONCLUSIONS

We present our results from experiments held during last two total solar eclipses, observed by our team (2017 August 21 and 2019 July 2) on the connection between near ground wind properties and shadow bands pattern. We find a link between the average near ground wind speed and the frequency of the bands. This relation is also confirmed by the matching directions of propagation of the bands and the wind before and after the totality of the eclipse from 2019.

These primary results support the hypothesis that the ground level wind's atmospheric scintillation may influence the pattern of the shadow bands.

Acknowledgements

This study is part of the project "Research on active solar processes during and beside total solar eclipses", funded by the National Science Fund of Bulgaria with contract No. KP-06-H28/4 (8-Dec-2018) and is supported by the Ministry of Education and Science project No. 577/17.08.2018, and the joint project of cooperation between the IA NAO, BAS, and the Department of Astronomy, Faculty of Mathematics, University of Belgrade, Serbia.

References

Chambers G. F.: 1899, *The Story of Eclipses*, George Newnes, Ltd., London, p. 46.

Codona J. L.: 1986, *Astronomy and Astrophysics*, 164(2), 415–427.

Guillermier P., Koutchmy S.: 1999. *Total Eclipses: Science, Observations, Myths and Legends*, Springer Publishing, p. 151.

Hastings C. S.: 1924, The shadow bands of total eclipses, *Popular Astronomy*, 32, 411.

Littmann M., Willcox K., Espenak F.: 1999, *Totality: eclipses of the Sun (2nd edition).* Oxford University Press. p. 119.

Marschall L. A., Mahon R., Henry R. C.: 1984, Observations of shadow bands at the total solar eclipse of 16 February 1980. *Applied Optics*, 23, 4390–4393.

Standord A. L.: 1973, On Shadow Bands Accompanying Total Solar Eclipses, *American Journal of Physics*, 41, 731.

Proceedings of the XII Serbian-Bulgarian Astronomical Conference
(XII SBAC) Sokobanja, Serbia, September 25-29, 2020
Editors: L. Č. Popović, V. A. Srećković, M. S. Dimitrijević and A. Kovačević
Publ. Astron. Soc. "Rudjer Bošković" No 20, 2020, 85-91

SOLITONS IN THE IONOSPHERE – ADVANTAGES AND PERSPECTIVES

MIROSLAVA VUKCEVIC and LUKA Č. POPOVIĆ

Astronomical Observatory, Volgina 7, 11060 Belgrade, Serbia
E-mail: mvukcevic@aob.rs, lpopovic@aob.rs

Abstract. We present the recent work on the soliton formation possibility within the ionospheric layers of the Earth. Linear waves are investigated very well and their existence is confirmed in the huge literature. However, they are unable to explain some of the fine structures observed in the Earth's atmosphere and ionosphere. Detection of the ion/electron density drop and consequently the density drop of the neutral gas is difficult to explain by linear theory. Therefore, we employ nonlinear magneto-hydro-dynamic (MHD) theory, investigating the perturbation of compressible fluid under the influence of self-gravity, rotation and magnetic field in the plain geometry. Solution that we search for is soliton, stable wave of constant amplitude and group velocity. Such a solution is more accurate and its space and time localization give an opportunity for instant detection. We have derived necessary condition for the vortex type of the solution as a balance of dispersive and nonlinear effects. At higher latitudes dispersion is mainly driven by rotation while near the Equator magnetic field modifies the solution within the E and F layer. This very general description of the ionosphere provides the conclusion that the unperturbed layer thickness cannot be taken as an ad hoc assumption, it is rather a consequence of the equilibrium property.

1. INTRODUCTION

Solitons are solution of nonlinear equations, in general. There are one dimensional solitons (ocean surface gravity waves, spiral density waves) and two-dimensional solitons (vortex, hurricane). They are stationary solutions as a balance between nonlinearity and dispersion. It can be concluded from the linear dispersion relation on the dispersive properties of the system and it can give a hint for the type of integrable nonlinear equation. Dispersive properties are consequence of the frequency dependence on the wave number, meaning that different wave numbers are traveling with different velocities. It results in the fact that the top of the wave travels faster that the bottom, braking the wave as it is shown in Fig. 1.

Figure 1: Dispersive one-dimensional ocean surface wave.

Under certain circumstances, this dispersion is balanced by the nonlinear effects and that circumstances are defined by equilibrium values. If the soliton is possible to be created than these equilibrium parameters could be used to derive the size and amplitude of the structure.

In the case of ionosphere we derive the nonlinear equation that has two type of nonlinearity: vector and scalar. Vector term is responsible for some turbulent dynamics known as Hasegawa-Mima equation and Rossby waves. Scalar term will be responsible for the solitary vortex creation (Vukcevic & Popovic 2020; Vukcevic, 2019).

2. IONOSPHERE PROPERTIES

Ionosphere is divided in three layers, in general. D layer, located at 50-100 km from the Earth's surface, charged particles contribution can be neglected so that ponderomotive force effects are small compared to Coriolis force effects (Gershman, 1974). E layer is located at 100-150 km and ponderomotive force is on the order of the Coriolis one. F layer is at heights of 150-400 km with the dominant ponderomotive force.

We have used the following plasma conditions within the ionosphere: ions are unmagnetized so the ion velocity is the velocity of the neutral gas; ion velocity across the magnetic field is equivalent to gas velocity; ions are dragged by neutral gas completely while electrons are magnetized and frozen in external magnetic field; electron velocity is defined by ExB (Kaladze et al., 2004); viscous effects are neglected due to high Hartmann number (Kaladze et al., 2004).

Local coordinate system of the ionosphere layer is defined as in Fig. 2.

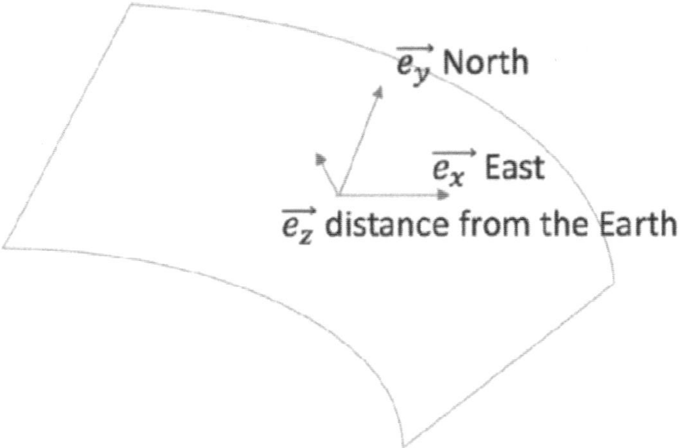

Figure 2: Local coordinate system of the ionosphere layer. In the horizontal plane, there are two axes x and y. Along the z axis is defined the distance of the horizontal plane from the Earth's surface.

Earth's angular velocity has the following components: $\Omega=\Omega(0,\sqrt{1-b^2},b\Omega)$, where the Equator is defined by b=0, while the pole is defined by b=1. Geomagnetic field is assumed to be magnetic dipole with components: $B_0=B_0(0,\sqrt{1-b^2},-2bB_0)$.

3. NONLINEAR EQUATION AND SOLUTION

Using set of standard fluid equations (continuity equation, momentum equation accompanied with Poisson's equation), non-constant thickness of the layer and drift approximation, we have derived following nonlinear equation

$$-0.2uf\frac{\partial}{\partial y}(B\phi - A\nabla^2\phi) - \frac{1}{2}B'\frac{\partial}{\partial y}\phi^2 + A(\nabla\phi\times\nabla)_z \quad \nabla^2\phi + +(\phi_0' B - \sigma_0')\frac{\partial}{\partial y}\phi - \phi_0' A\frac{\partial}{\partial y}\nabla^2\phi = 0, \tag{1}$$

with relevant frequencies

$$f = f_R + f_H = 2b\left(\Omega + \frac{enB_0}{\rho}\right)[v_x, v_y, 0)] \quad - \frac{2enbB_0}{\rho}\left[5(1-b^2)v_x, (1-2b^2)v_y, 0)\right] \quad . \tag{2}$$

Normalization of the variables are done by $2\Omega+H$ where $H=enB_0/\rho$. Here n is number of ionized particles while ρ is density of neutral gas.

Solution of the nonlinear equation has a form

$$\phi = \frac{2\lambda}{\nu}F(R), \tag{3}$$

where R=$\sqrt{\lambda}r$ is dimensionless radius of the structure in the moving frame, and F is the solution of following equation (Zakharov, Kuznetsov; 1974)

$$F = 2.4\left(\cosh\left(\frac{3}{4}R\right)\right)^{-\frac{4}{3}}. \tag{4}$$

Parameters

$$\lambda(x) = \frac{1}{A}\left(B - \frac{\sigma_0'}{u}\right), \quad \nu(x) = \frac{1}{2u}\left(\lambda A' + \frac{\sigma_0''}{u}\right), \tag{5}$$

are defined by gradient and second derivative of density with respect to x.
Potential has a form of solitary vortex traveling along y coordinate (northward) with constant velocity u.
If the normalization factor is not symmetric, solution is

$$F = \cosh\left(\frac{3}{4}R(1 + f(x)\Lambda)\right)^{-\frac{4}{3}}. \tag{6}$$

Soliton vortex is asymmetric elongated either along x or along y axis, depending on the frequency.

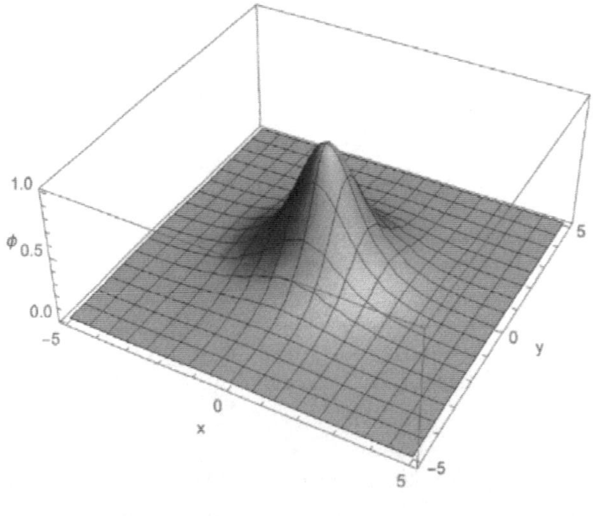

Figure 3: Symmetric solution of the potential; the same as it is shown in Fig. 2 presented in Vukcevic & Popovic (2020).

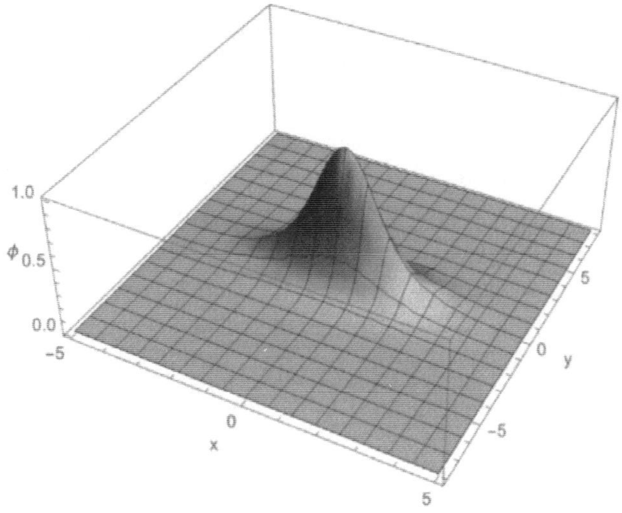

Figure 4: Asymmetric solution of the potential close to pole elongated along x axis; the same as it is shown in Fig. 3 presented in Vukcevic & Popovic (2020).

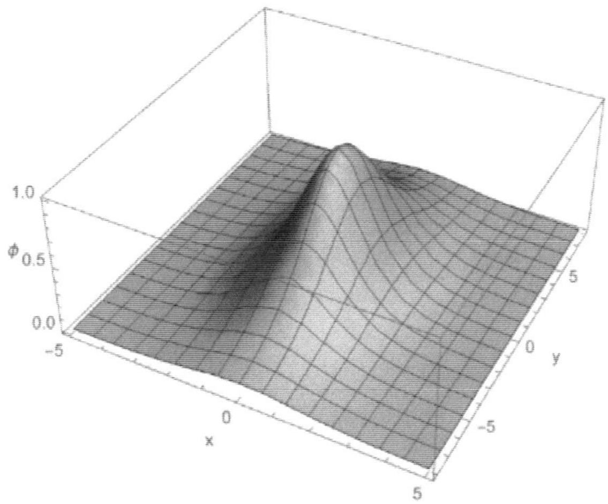

Figure 5: Asymmetric solution of the potential in the Equator vicinity elongated along y axis; the same as it is shown in Fig. 4 presented in Vukcevic & Popovic (2020).

4. RESULTS

Ionosphere D layer. Within this region, located 50– 100 km from the Earth's surface, we can assume that the contribution of charged particles can be neglected, so the ponderomotive force effects are small compared to the Coriolis force effects (Gershman, 1974). This means that it is likely to expect a solitary structure for small latitudes, close to the Equator, elongated along the y coordinate, while for high latitudes and at the pole, the soliton is symmetric, the size of the soliton will depend on the density gradient, and its velocity is normalized by $f=f_R=\Omega$.

Ionosphere E layer. This layer is located 100–150 km from the Earth's surface, and one can expect the creation of a soliton at all latitudes higher than 6^0, since the ponderomotive force is on the order of the Coriolis one. In this case, soliton velocity is defined by $f = 2(\Omega + H)$ at the pole and for latitudes close to the pole. Since the value H has the opposite sign to Ω, the cancelation of the vortex structure is possible when these two terms are on the same order or it is possible to change the moving direction of the soliton structure. Next, the size of the soliton is defined by R and, consequently, by soliton velocity u, which for this case is defined by $2(\Omega + H)$; one expects the size to increase com- pared with the same case for the D layer. As far as the low-latitude case is concerned, the latitudes close to the Equator and the size and velocity of the soliton are dependent on the value H compared to Ω, and even more, the soliton is not symmetric but rather extended along the y axes, since $f=f(x,y)$.

Ionosphere F layer. The ionospheric F layer is for heights 150–400km from the Earth's surface. In this case, for all latitudes soliton structure is mainly defined by the value 0.2H. At the pole, the soliton is symmetric and velocity is defined by $f = 2H$, while close to the Equator one can expect a soliton elongated along the x axis, but moving in the opposite direction compared to the E and D layers, since $f = f(x,y)=f_H$.

5. CONCLUSIONS

A series of direct observations of such soliton structures are carried out either from the Earth's surface or onboard the satellites. We have summarized all possible soliton structure formations at different latitudes, as well as at different ionospheric layers. The soliton size and velocity are constant but defined by different values of ionospheric parameters.

We hope that this model will be used in explanations of the ionosphere structures as well as in testing the physics background of complex ionosphere simulations. This model can be used not only to model the ionosphere structure, but also for different astrophysical systems, e.g., accretion disks, where the thickness effects could be very important. Therefore, finite thickness effects should be taken into account. However, this approach can be improved by trying to find out the correlation between soliton structure dynamics and other methods used to identify the ionospheric anomalies. Also, it would be of great importance to investigate the stability of the soliton structure as the subject of small disturbances and apply it to the study of the interaction between the solitons within different ionospheric layers. All of these mentioned issues will be considered in further research.

References

Gershman B. N.: 1974, *Dynamics of Ionospheric Plasma*, Nauka, Moscow

Kaladze T. D., Aburjania, G. D., Kharshiladze, O. A., Horton, W., Kim, Y.-H: 2004, *J. Geophys. Res.*, 109, A05302

Vukcevic M.: 2019 *MNRAS*, 484, 3410-3418

Vukcevic M., Popović Č. L.: 2020, *Nonlinear Processes in Geophysics*, 27, 295-306

Zakharov V. E., Kuznetsov E. A.: 1974, *Sov. Phys. JETP*, 39, 285–289

Proceedings of the XII Serbian-Bulgarian Astronomical Conference
(XII SBAC) Sokobanja, Serbia, September 25-29, 2020
Editors: L. Č. Popović, V. A. Srećković, M. S. Dimitrijević and A. Kovačević
Publ. Astron. Soc. "Rudjer Bošković" No 20, 2020, 93-105

STORM ACTIVITY OVER BALKAN REGION DURING MAY 2009

ALEKSANDRA KOLARSKI

Technical Faculty "Mihajlo Pupin", University of Novi Sad,
23000 Zrenjanin, Serbia
E-mail: aleksandrakolarski@gmail.com

Abstract. Intense storm activity over Balkans (40°/48° N, 12°/23° E) at the end of May 2009 was analyzed. Surveying was carried out by integration of satellite and ground-based observations. Very Low Frequency (VLF) signals (3-30 kHz) recorded by Absolute Phase and Amplitude Logger station in Belgrade (44.85° N, 20.38° E), video recordings of sprite events from ITALIAN METEOR and TLE NETWORK and lightning stroke data from European Cooperation for Lightning Detection network were inspected for possible relationship. Different type and magnitude of perturbations on monitored VLF signals were observed, even originated from same lightning discharge. Correspondence between all three examined phenomena was found, in some of analyzed cases.

1. INTRODUCTION

Although it has been investigated for many decades and from numerous different aspects, thunderstorm activity as scientific area is still under active interdisciplinary research. In recent years, thunderstorm activity is increasingly interpreted as terrestrial hazard (equally to earthquakes, landslides, tsunamis, forest fires, snow blizzard etc.), especially taking into account accompanying floods, destructive winds, damage and deaths from lightning strokes. Scientific interest in thunderstorm ranges from topics that cover indirect effects on human health (Elliot et al. 2014), to climatological studies (Cliverd et al. 2017, Romps et al. 2014, Finney et al. 2018). Physical processes involved in thunderstorm activity cover many still opened questions, broadly and diversely actively researched. Only some of them are addressed to processes of thunderstorm electrification, initiation mechanism of lightning leaders, causative mechanisms between thunderstorms and induced lower ionospheric perturbations (relatively recent results can be found in e.g. Silber and Price (2017, and references therein).

Changes in climate during recent years are more easily noticeable and more increasingly recognizable in Balkan regions. Although Balkans are still not considered as typical severe weather regions, extreme weather events like floods and landslides in 2014 (Stadtherr et al. 2016, Abolmasov et al. 2017, Djurić et al. 2017, Suto et al. 2016) and heat and cold wave in 2017 (Kew et al. 2019, Anagnostopoulou et al., 2017), suggest that soon in future, extreme weather could become habitual feature over these territories, too.

Extreme weather monitoring over Balkans, is still rare and sporadic and not much is known about electrical properties of such intense storm systems (Kolarski 2019, Kolarski 2020). To my knowledge, Balkan region in wider sense (including Adriatic coastal region both Italian and Croatian, then geographical territories of Bosnia, Croatia and Serbia and other former Yugoslavia republics) was never systematically studied in terms of atmospheric electrical properties related to storm activity and Transient Luminous Event (TLE) observations (Arnone et al., 2019). Intense storm activity from end of May 2009, that hit region of Adriatic coast and Balkan Peninsula inland, represents the unique opportunity to get insight in electrical properties and underlying mechanisms of thunderstorm systems generated over Balkans, through integrated satellite and ground-based observations.

2. EXPERIMENTAL SET-UP AND OBSERVATIONS

The survey of data from four independent sources was carried out and included data from EUMETSAT (European Organization for the Exploitation of Meteorological Satellites (http://www.eumetsat.int/)) weather satellite, data set about registered atmospheric discharges obtained from EUCLID (European Cooperation for Lightning Detection http://www.euclid.org/)) Network, data set about optically documented TLE events obtained from IMTN (ITALIAN METEOR and TLE NETWORK (http://www.imtn.it/)) network, and last data set was about Very Low Frequency (VLF) radio signal records from AbsPAL (Absolute Phase and Amplitude Logger) receiving system located in Belgrade (44.85° N; 20.38° E), from Institute of Physics' database. The presented analysis takes the VLF signal amplitude and phase delay data as the basic data set related to two other ground-based datasets. VLF (3-30 kHz) signal records, video records of sprite events and detected lightning strokes data were analyzed in detail, in order to find coincidence and possible relationship between these three phenomena during the stormy night 27th – 28th of May, 2009, while satellite data were only qualitatively analyzed.

EUMETSAT satellite documented 6 hours of intense storm activity over Central and Southeast Europe, from 9 o'clock p.m. on 27th to 3 o'clock a.m. on 28th of May 2009. During this stormy night, the core of the storm activity that occured and started over northern italian adraiatic cost, was moving eastward to and along Dalmatian adriatic coastal region and then further towards and into the Balkan Peninsula inland. On qualitatively analyzed satellite graphic material, storm activity is rounded by generalized frames and plotted in different colours, from black to blue, on each of the hourly satellite shots. The first and the last frames from satellite graphic material, with storm activity rounded in pink, are given in

Fig 1. Generalised movement and enlarged position of the storm system's core with legend of colours used are given in Figs. 2 and 3, respectively.

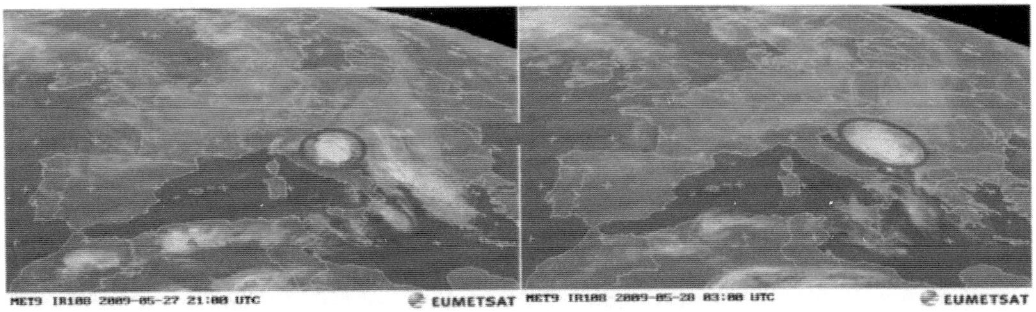

Figure 1: EUMETSAT weather satellite data registrations over Central and Southeast Europe during the stormy night 27th – 28th of May, 2009, first and last hourly frames.

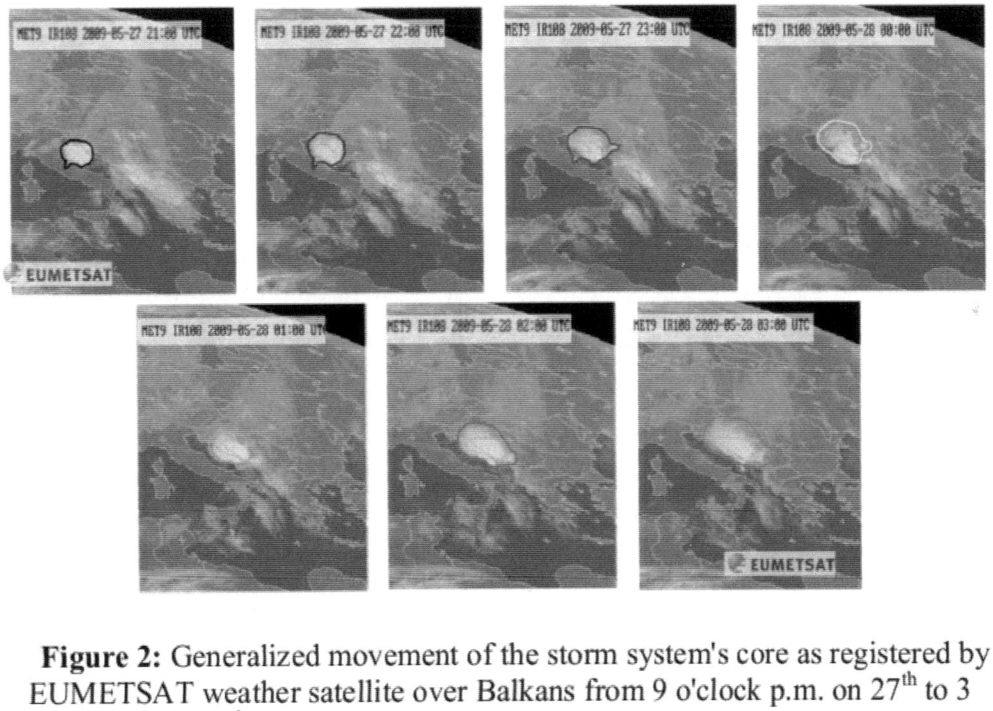

Figure 2: Generalized movement of the storm system's core as registered by EUMETSAT weather satellite over Balkans from 9 o'clock p.m. on 27th to 3 o'clock a.m. on 28th of May 2009 during the stormy night of May 27 – 28, 2009.

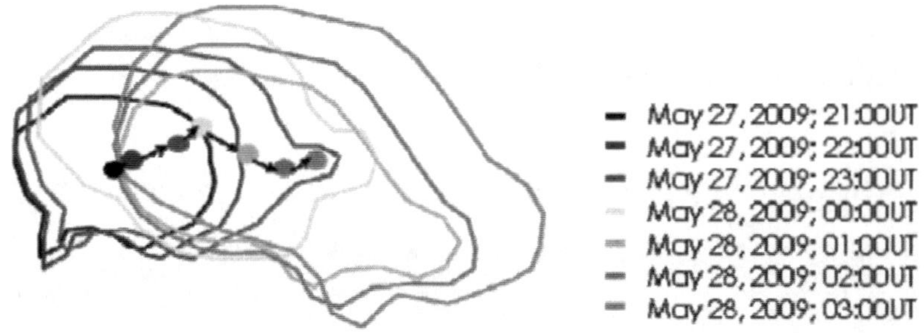

a) b)

Figure 3: a) Generalized position and movement of the storm system's core as registered by EUMETSAT weather satellite over Balkans during the stormy night 27th – 28th of May, 2009 and b) legend of colours used

The second data set analyzed was about lightning stroke events. For this particular night, over the observed geographical area enclosed within 12 to 23 degrees east in longitude and 40 to 48 degrees north in latitude, in period of 6 hours, EUCLID network reported intense storm activity with almost 22000 lightning stoke events, both of cloud-to-ground (CG) and inter-cloud (IC) type. Peak current distribution regarding the type of strokes and their peak current intensity and polarity is given in Fig. 4, while as projected on Earth's surface is given in Fig. 5. Stroke event distribution regarding the number of reported events by type is given in Fig. 6. The predominant are stroke events with negative polarity and with relatively small peak currents, which accounts for more than 2/3 of total strokes reported. The predominant are CG type of stroke events, which accounts for about 98% of total strokes reported, and among them those with negative polarity and with relatively small peak currents. In this data set there were only about 2% stroke events of IC type, with about 300 strokes, with similar peak current intensities.

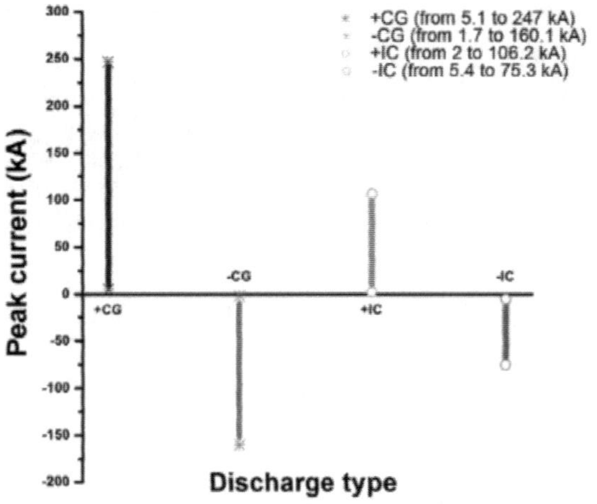

Figure 4: Peak currents distribution with peak currents from 160 kA (–CG) to 247 kA (+CG), as reported by EUCLID network.

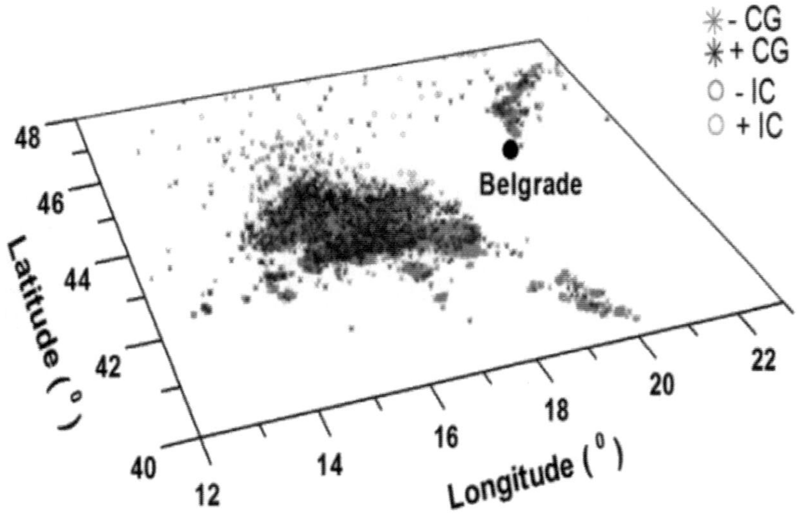

Figure 5: Stroke event distribution – projection on Earth's surface, as reported by EUCLID network.

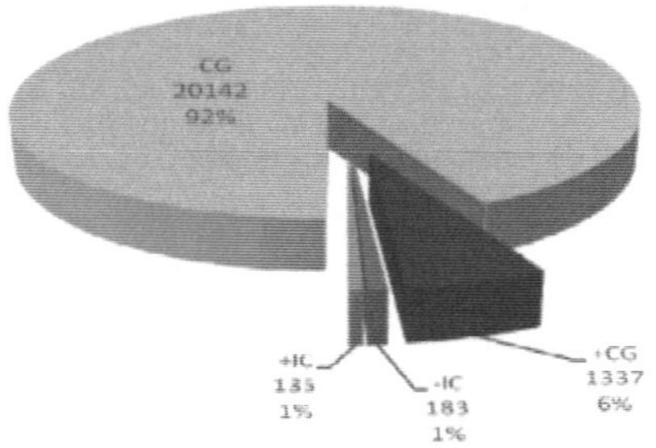

Figure 6: Stroke event distribution – number of reported events, as reported by EUCLID network.

The third data set analyzed was about TLE events. During this particular night, by IMTN network within 6 hours of intense storm activity, several dozens of TLE events were reported and optically documented from Ferrara station (44.82° N, 11.62° E) UFOcaptureV2 camera (directed with az. 150.49° and ev. 24.57° and with AOVs 122.8° and 97.1° horizontal and vertical, respectively, looking from the 45° N towards SE), oriented SE and with FOV covering area from Adriatic Sea and coast towards inner regions of Balkan Peninsula. According to obtained video material, within the area enclosed by 12-16 degrees east in longitude and 42-46 degrees north in latitude, there were 71 sprites and 10 halos in total documented between 21:42:11.4 UT on 27th and 02:18:55.0 UT on 28th of May 2009, with vast majority of the reported TLE activity located slightly over the horizon. Some of the sprite events reported by Ferrara station on night 27th – 28th of May, 2009 are given in Fig. 7, with sprite events rounded by yellow ellipse.

Schematic diagrams of locations, orientation and field of view of IMTN network cameras dedicated for observation of night-time phenomena such as meteors and TLE events, mostly over Italian land and sea, but also over surrounding areas including back of the Adriatic sea and further towards Balkans, which was particularly significant for this analysis, can be found at www.imtn.it (Fig. 8).

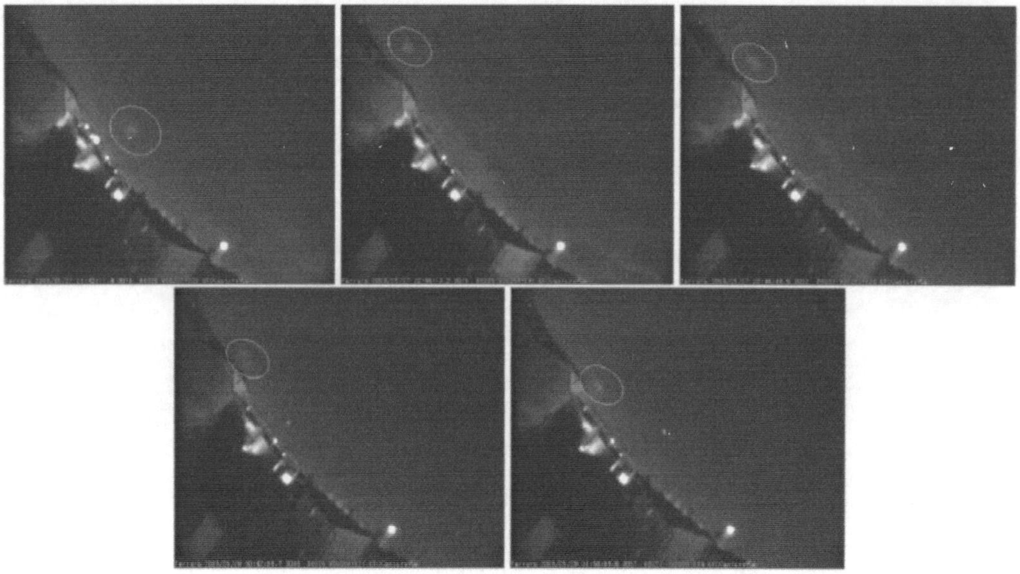

Figure 7: Some sprite events reported by Ferrara station on night 27th – 28th of May, 2009.

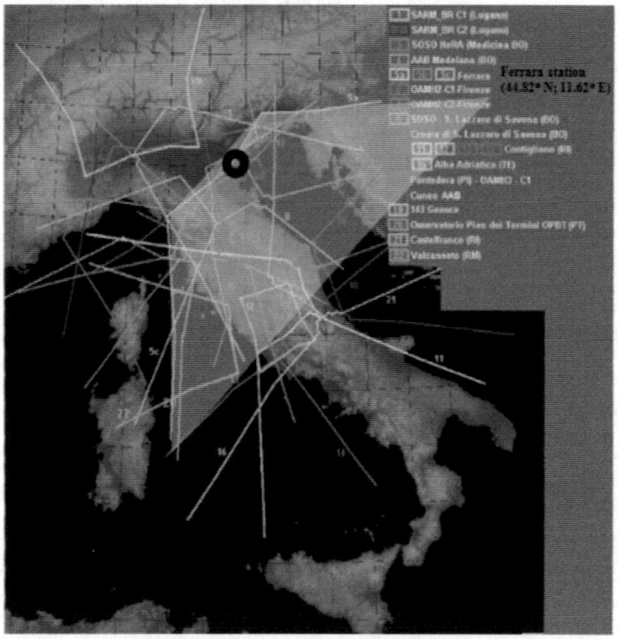

Figure 8: ITALIAN METEOR and TLE NETWORK (IMTN) with Ferrara station (black circle) with FOV (yellow polygon) covering area from Adriatic Sea and coast towards inner regions of Balkan Peninsula.

Within fourth dataset, propagation paths of VLF signals, emitted from different directions towards Belgrade AbsPAl receiver from transmitters located in different parts of the world, were analyzed. This data set was used as basic dataset. It is important to notice that Belgrade AbsPAl receiving system operates in stable mode from 2004, with first records made at the end of 2003. System contains 6 channels with 5 dedicated for VLF signals receiving and 1 reserved for system's time synchronization with GPS satellites. During the stormy night 27th –28th of May, 2009, these 5 signals were recorded: signal from USA with code name NAA emitted on frequency 24.0 kHz, signal from UK with code name GQD emitted on frequency 22.1 kHz, signal from Australia with code name NWC emitted on frequency 19.8 kHz, signal from Germany with code name DHO emitted on frequency 23.4 kHz and signal from France with code name FTA emitted on frequency 20.9 kHz. Characteristics of VLF signal transmitters are given in Table 1.Three propagation paths of VLF signals transmitted from USA, UK and Germany towards Belgrade receiver were analyzed in detail (bold in Table 1) and are given in Fig. 9. Signal emitted from USA – NAA/24.0 kHz, with Great Circle Path (GCP) distance of about 6.5 Mm is partly over-land and partly over-sea signal trace, with most of the path over sea. Signal emitted from UK – GQD/ 22.1 kHz, with GCP distance of about 2 Mm is partly over-land and partly over-sea signal trace, with most of the path over land. Signal emitted from Germany – DHO/23.4 kHz, with GCP distance of about 1.3 Mm has over-land path.

Table 1: List of VLF transmitters

VLF signal code and frequency	Transmitter location	Emitted power	GCP distance
NAA/24.0 kHz	**Maine, USA (44.63 N; 67.28 W)**	**1000 kW**	**6547 km**
GQD/22.1 kHz	**Skelton, UK (54.72 N; 2.88 W)**	**500 kW**	**1982 km**
NWC/19.8 kHz	H. E. Holt, Australia (27.2 S; 114.98 E)	1000 kW	11975 km
DHO/23.4 kHz	**Rhauderfehn, Germany (53.08 N; 7.62 E)**	**800 kW**	**1301 km**
FTA/20.9 kHz	Sainte-Assise, France (48.54 N; 2.58 E)	400 kW	1413 km

Figure 9: VLF signals' GCPs as registered by Belgrade AbsPAL receiver system located at the Institute of Physics (44.85° N; 20.38° E), in Serbia, on night 27th – 28th of May, 2009.

Integrated analyzed data from all ground-based sources during this particular night, over area of interest are given in Fig. 10. VLF signal traces, recorded by

Belgrade AbsPAL receiver are presented with thick black solid lines. Stroke events reported by EUCLID network, within 2-minute time intervals enclosing each of TLEs optically documented by IMTN network, are presented by black squares. Ferrara station camera's FOV is presented with thin dashed black line. Area enclosing all TLEs optically documented by IMTN network is presented with white dash-dotted somewhat thicker line. Analyzed VLF signals, coming from left, pass over region that was hit by the storm activity reported by EUMETSAT. It should be kept in mind that this graph is projection and because the Earth's surface is curved, these signal traces are even closer to each other, especially near Belgrade receiver.

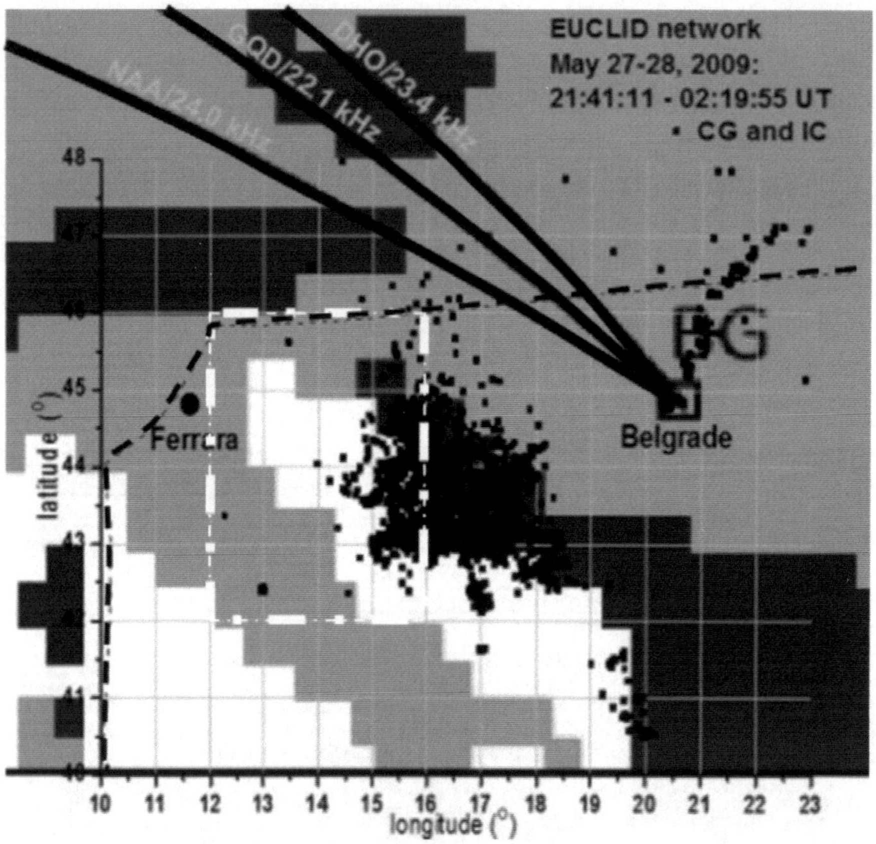

Figure 10: Integrated data from ground-based observations on night 27th – 28th of May, 2009: analyzed VLF signals received in Belgrade (thick solid black lines), strokes reported by EUCLID network (filled small black squares) and TLEs documented by IMTN network (within area enclosed by white rectangular of dash-dotted somewhat thicker line) (Ferrara station (44.82° N; 11.62° E) presented by black filled circle with FOW presented by thin dashed black line and Belgrade AbsPAL receiver (44.85° N; 20.38° E) presented by big hollow black square).

3. RESULTS AND DISCUSSION

Lower Ionosphere, as the lowest ionospheric region, extends in height from 50 to 90 km in altitude range, overlapping with atmospheric regions of mesosphere and lower thermosphere. Earth-Ionosphere waveguide is limited with Earth's surface at its lower boundary and lower ionospheric D-region (50-90 km) lower boundary, at its upper limit. Employing VLF radio signals for exploration of the lower Ionosphere, nowadays is widely used remote sensing technique. VLF radio signals, globally propagating through Earth-Ionosphere waveguide, are dependent on electron density changes that take place in the lower ionosphere, as induced by variety of phenomena from extraterrestrial to terrestrial origin. Ionospheric conductivity perturbations caused by lightning activity, as one of the agents of terrestrial origin, usually produce VLF signal amplitude and phase delay disturbances of small amount and of duration on time scale from several ms to several tens of ms, sometimes at levels of VLF signal noise, but in some cases can produce significant striking long-lasting VLF signal disturbances (detailed analysis is beyond the scope of this paper, form more details see e.g. Silber and Price 2017, and references therein).

In conducted analysis, focus was on ionization changes along the propagation path of VLF radio signals induced by the strong release of energy by atmospheric lightning discharges. The increased ionization is apparent in the perturbation of the signal amplitude and phase delay with respect to regular undisturbed ionospheric conditions. The perturbations can manifest themselves thorough increase or decrease, or as complex - both increase and decrease of VLF signal Amplitude and Phase delay. Examples of different propagation conditions within Earth-Ionosphere waveguide, within 1min time intervals, during May 2009 are given in Fig. 11. On the left panel, the example of unperturbed propagation conditions on 2nd of May is given. The typical example of perturbed propagation conditions, with isolated VLF perturbation from night 27th – 28th of May is shown on the middle panel. The typical example of severely perturbed propagation conditions with numerous VLF perturbations characteristic for intense storm activity during night 27th – 28th of May is given on right panel. The effects of induced ionization changes have been observed along different propagation paths during the entire night 27th –28th of May, 2009, but the idea was to inspect in detail narow time intervals in which the sprite events were reported and optically documented by cameras. On the other hand, VLF traces with obvious, strong and clear perturbations such as typically on NAA and GQD signals were chosen for further detailed analysis.

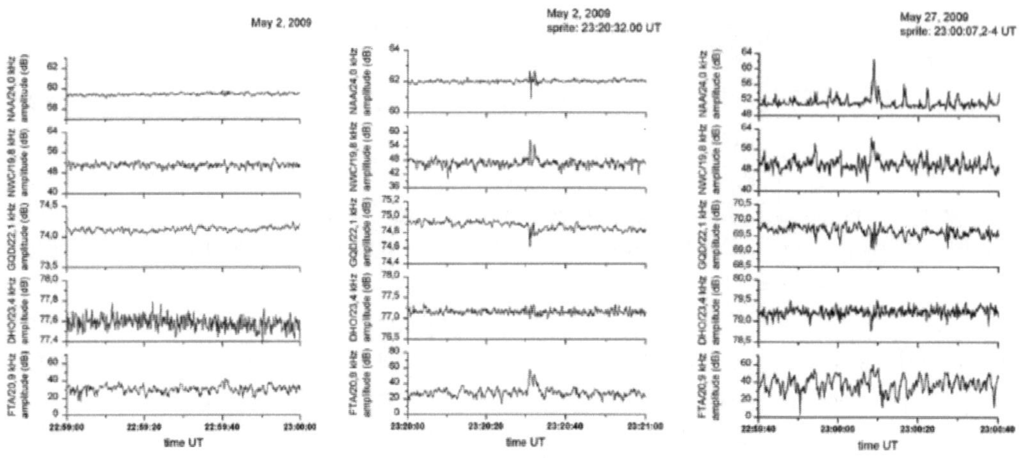

a) b) c)

Figure 11: Different radiopropagation conditions within Earth-ionosphere waveguide during May 2009: a) unperturbed propagation conditions, b) perturbed propagation conditions with isolated VLF perturbation and c) severly perturbed propagation conditions.

Typical examples of VLF perturbations during this stormy night are given in Fig. 12 (denoted by red arrows). Two-minute time intervals enclose sprite events reported by IMTN network. Reported sprite events, VLF perturbation and reported stroke events, all are in very narrow time interval, so they belong to the same storm activity. Accompanying lightning stroke events for above mentioned two sprite cases, within observed arrea of interest (40° - 48° N; 12° - 23° E) as reported by EUCLID network, are given in Fig. 13, presented with black squares.

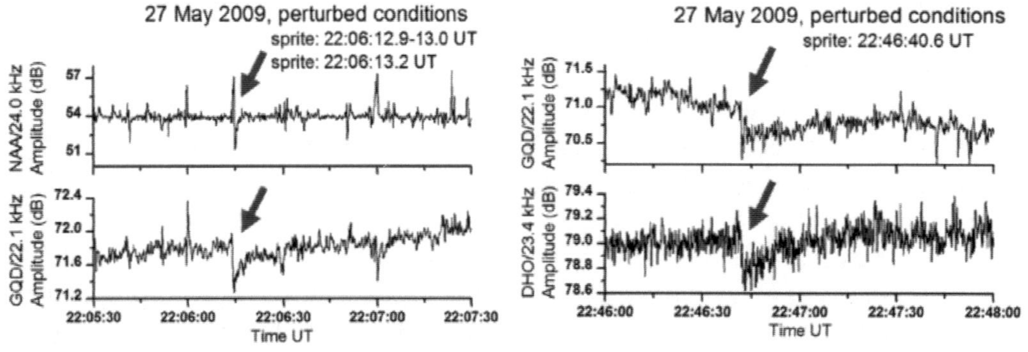

a) b)

Figure 12: Perturbed propagation conditions on analyzed VLF signals recorded in Belgrade during the night 27th –28th of May, 2009; typical examples of VLF perturbations as denoted by red arrows.

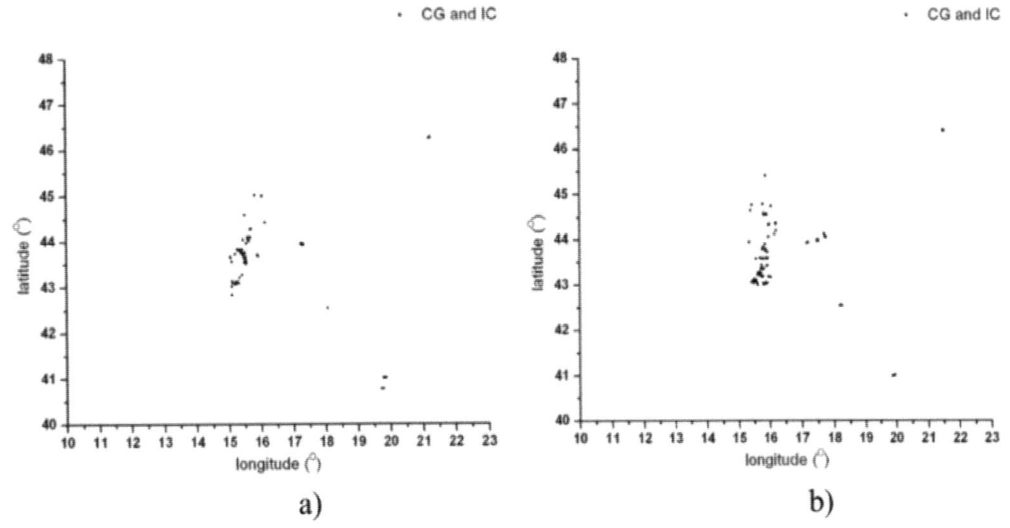

Figure 13: Lightning strokes reported by EUCLID network, over area 40° - 48° N and 10° - 23° E, within 2-minute time intervals enclosing sprite events reported by IMTN network given in examples from Figure 12.

During this entire night, on all considered VLF traces, series of amplitude and phase delay perturbations were recorded. Summaries, that can be applied to all 81 cases analyzed, that were related to optically documented TLEs, are as follows:

- VLF perturbations often appeared simultaneously on all analyzed signal traces,

- although related to the same discharge event, VLF perturbations manifested themselves different in type and magnitude, which was attributed to the relative perturbed region's distance from VLF signal's GCP.

- in all 2-min time samples that enclosed each of TLE events optically documented, registered VLF perturbation corresponded to TLE event within intervals of few hundreds of ms. Even though VLF perturbation and sprite event did not exactly coincide within inspected time samples, these phenomena originated from the same storm activity.

- causative CG strokes were not found for all documented sprites. In most cases, lightning strokes were reported over observed area within inspected 2-min time samples and correspondence between two groups of phenomena, like VLF perturbations and CG strokes on one and VLF perturbations and TLE events on the other hand, was found. In some cases, correspondence between all these three types of events, was found.

- one-to-one corresponding between VLF perturbations, TLEs and lightning strokes was not found in all analyzed cases. Possible reasons are that stroke events were simply missed by EUCLID, or that sprite was not captured by IMTN camera, or that VLF perturbation might not is caused by sprite.

4. CONCLUSIONS

During the same storm activity which may include a number of strokes in narrow time interval of few hundreds of ms following types of events were noticed:

a) stroke is followed by sprite and sprite was followed by VLF perturbation,

b) initial stroke for sprite was registered and the sprite preceded or was observed during the VLF perturbation,

c) initial stroke for sprite was not registered and the sprite preceded or was observed during the VLF perturbation.

It can be concluded:

- in case a) the VLF perturbation is caused by scattering on the sprite body.
- in case b) if the sprite is preceding the VLF perturbation, the later is caused by scattering on the sprite body; however, if the sprite is observed during the VLF perturbation, the later is caused by electron density changes in waveguide due to the stroke. The appearance of the sprite during the VLF perturbation can prolong VLF perturbation duration.
- in case c) the cause of sprite and VLF perturbation was not detected.

In cases with simultaneous VLF perturbations registrations, reasonable assumption is proximity between perturbed region and receiver. During night 27th –28th of May, 2009, intense storm activity was reported both by satellite and ground-based sources, over area few hundreds of km (up to 400 km) away from Belgrade receiver. Since all analyzed VLF signal traces passed over very close to region hit by reported intense storm activity and in vicinity of the area where TLEs were documented, it is not likely that VLF perturbations were induced by some other storm activity originated outside this region.

Acknowledgements

Author thank to the funder Ministry of Education, Science and Technological Development of Republic of Serbia, to Institute of Physics, University of Belgrade, Serbia, for providing the VLF data and to D. Šulić for VLF instrumental set-up. Author thank to all third parties that provided data that have been used in this research: IMTN network for providing the optical data on the TLE events, especially F. Zanotti from Ferrara station, Italy and J. Bór from the Geodetic and Geophysical Institute, Research Centre for Astronomy and Earth Sciences, Hungarian Academy of Sciences, Sopron, Hungary, EUCLID network and especially G. Milev from Elektroinštitut Milan Vidmar, Ljubljana, Slovenia, for providing the data on the lightning strokes and EUMETSAT for graphic material.

References

Abolmasov B., Marjanović M., Djuric U., Krušić J., Andrejev K.: 2017, Massive Landsliding in Serbia Following Cyclone Tamara in May 2014 (IPL-210), *4th World Landslide Forum Ljubljana, Slovenia, Workshop on World Landslide Forum*, 473-484.

Anagnostopoulou C., Tolika K., Lazoglou G., Maheras P.: 2017, The Exceptionally Cold January of 2017 over the Balkan Peninsula: A Climatological and Synoptic Analysis, *Atmosphere 2017*, 8(12), 252,

Arnone E., Bór J., Chanrion O., Barta V., Dietrich S., Enell C-F., Farges T., Füllekrug M., Kero A., Labanti R., Mäkelä A., Mezuman K., Odzimek A., Popek M., Prevedelli M., Ridolfi M., Soula S., Valeri D., Van der Velde O., Neubert T.: 2019, Climatology of Transient Luminous Events and Lightning Observed Above Europe and the Mediterranean Sea, *Surveys in Geophysics,* 41, 10.

Clilverd M. A., Duthie R., Rodger C. J., Hardman R. L., Yearby K. H.: 2017, Long-term climate change in the D-region, *Scientific Reports*, 7(1), 16683.

Đurić D., Mladenović A., Pešić-Georgiadis M., Marjanović D. M., Abolmasov B.: 2017, Using multiresolution and multitemporal satellite data for post-disaster landslide inventory in the Republic of Serbia, *Landslides,* 14, 1467–1482.

Elliot A. J., Hughes H. E., Hughes T. C., Locker T. E., Brown R., Sarran C., Clewlow Y., Murray V., Bone A., Catchpole M., McCloskey B., Smith G. E.: 2014, The impact of thunderstorm asthma on emergency department attendances across London during July 2013, *Emerg Med J*, 31(8), 675-678.

Finney D. L., Doherty R. M., Wild O., Stevenson D. S., MacKenzie I. A., Blyth A. M.: 2018, A projected decrease in lightning under climate change, *Nature Climate Change*, 8, 210-213.

Kew S. F., Philip S. Y., van Oldenborgh G. J., van der Schrier G., Otto F. E. L., Vautard R.: 2019, The Exceptional Summer Heat Wave in Southern Europe 2017, *Bull. Amer. Meteor. Soc.*, 100, S49–S53.

Kolarski A.: 2019, Atmospheric disturbances due to severe stormy weather, Book of Abstracts, "Integrations of satellite and ground-based observations and multi-disciplinarity in research and prediction of different types of hazards in Solar system", Petnica, Science Center, May 10-13, 2019, Valjevo, Serbia, Geographical Institute "Jovan Cvijić" SASA, Belgrade, Serbia, p. 42.

Kolarski A.: 2020, Storm activity over Balkan region during may 2009, Book of Abstracts, "XII Serbian-Bulgarian Astronomical Conference (XII SBAC)" September 25-29, 2020, Sokobanja, Serbia, Astronomical Observatory, Belgrade, Serbia, p. 75.

Romps D. M., Seeley J. T., Vollaro D., Molinari J.: 2014, Projected increase in lightning strikes in the United States due to global warming, *Science*, 346(6211), 851_854.

Silber I., Price C.: 2017, On the Use of VLF Narrowband Measurements to Study the Lower Ionosphere and the Mesosphere–Lower Thermosphere, *Surveys in Geophysics*, 38(2), 407–441.

Stadtherr L., Coumou D., Petoukhov V., Petri S., Rahmstorf S.: 2016, Record Balkan floods of 2014 linked to planetary wave resonance, *Science Advances*, 2, e1501428.

Suto K., Urosevic M., Komatina S.: 2016, An MASW survey for landslide risk assessment: A case study in Valjevo, Serbia, *Chang Mai Journal of Science*, 43 (6 Special Issue 2), 1249-1258.

Proceedings of the XII Serbian-Bulgarian Astronomical Conference
(XII SBAC) Sokobanja, Serbia, September 25-29, 2020
Editors: L. Č. Popović, V. A. Srećković, M. S. Dimitrijević and A. Kovačević
Publ. Astron. Soc. "Rudjer Bošković" No 20, 2020, 107-121

ACTIVITIES OF SERBIAN SCIENTISTS IN EUROPLANET

ALEKSANDRA NINA[1], MILAN RADOVANOVIĆ[2],
LUKA Č. POPOVIĆ[3], ANA ČERNOK[4], BRATISLAV P. MARINKOVIĆ[1],
VLADIMIR A. SREĆKOVIĆ[1], ANĐELKA KOVAČEVIĆ[5],
JELENA RADOVIĆ[6], VLADAN ČELEBONOVIĆ[1],
IVANA MILIĆ ŽITNIK[3], ZORAN MIJIĆ[1], NIKOLA VESELINOVIĆ[1],
ALEKSANDRA KOLARSKI[7] and ALENA ZDRAVKOVIĆ[8]

[1]*Institute of Physics Belgrade, University of Belgrade, Pregrevica 118, 11080 Belgrade, Serbia*
[2]*Geographical Institute Jovan Cvijić SASA, 11000 Belgrade, Serbia*
[3]*Astronomical Observatory, Volgina 7, 11060 Belgrade, Serbia*
[4]*Royal Ontario Museum, Center for the applied planetary mineralogy, Toronto, ON, Canada*
[5]*Department of astronomy, Faculty of Mathematics, University of Belgrade Studentski trg 16, 11000 Belgrade, Serbia*
[6]*Charles University, Faculty of Mathematics and Physics, Ke Karlovu 3, 12116, Prague 2*
[7]*Technical Faculty "Mihajlo Pupin", University of Novi Sad, 23000 Zrenjanin, Serbia*
[8]*Faculty of Mining and Geology, University of Belgrade, Belgrade*
E-mail: sandrast@ipb.ac.rs, bratislav.marinkovic@ipb.ac.rs, vlada@ipb.ac.rs, vladan@ipb.ac.rs, zoran.mijic@ipb.ac.rs, veselinovic@ipb.ac.rs
E-mail: m.radovanovic@gi.sanu.rs, lpopovic@aob.rs, ivana@aob.rs, acernok@rom.on.ca, andjelka@matf.bg.ac.rs, radovicj95@gmail.com, aleksandrakolarski@gmail.com, alena.zdravkovic@rgf.bg.ac.rs

Abstract. The Europlanet Society, an organization which promotes the advancement of European planetary science and related fields, has 10 hubs. The Serbian Europlanet Group (SEG) is included in the Europlanet South Eastern European Hub (ESEEH) and, currently, has 20 active scientists.

In this work, we present activities of SEG. Primarily, we describe two Europlanet workshops organized in the Petnica Science Center: "Geology and geophysics of the solar system bodies" and "Integrations of satellite and ground-based observations and multi-disciplinarity in research and prediction of different types of hazards in Solar system" that occurred in 2018 and 2019, respectively, and the Europlanet session during XII Serbian-

Bulgarian Astronomical Conference that occurred in Sokobanja 2020. In addition, we present other activities that were primarily aimed at connecting SEG members coming from six institutions as well as the promotion of the Europlanet and ESEEH organizations.

1. INTRODUCTION

The Europlanet society is an organization which promotes the European planetary science and related fields. Its aims are to support the development of planetary science at a national and regional level, particularly in countries and areas that are currently under-represented within the community, and early career researchers who establish their network within the Europlanet: the Europlanet Early Career (EPEC) network (https://www.europlanet-society.org/early-careers-network/).

Two Europlanet projects (the Europlanet 2020 Research Infrastructure and the Europlanet 2024 Research Infrastructure (RI)) are funded through the European Commission's Horizon 2020 programme. The first one, lasting 4 years, ended 2020, while the second one runs for four years from February 2020 until January 2024. The latest is led by the University of Kent, UK, and has 53 beneficiary institutions from 21 countries in Europe and around the world, with a further 44 affiliated partners. It provides free access to the world's largest collection of planetary simulation and analysis facilities, data services and tools, a ground-based observational network and programme of community support activities.

The Europlanet consists of 10 Regional Hubs:

- Benelux
- Central Europe: Austria, Czech Republic, Hungary, Poland, Slovenia and Slovakia
- France
- Germany
- Ireland and UK
- Italy
- Northern Europe: Denmark, Estonia, Finland, Iceland, Latvia, Lithuania, Norway and Sweden
- Southeast Europe: Bulgaria, Croatia, Cyprus, Greece, Romania, and Serbia
- Spain and Portugal
- Switzerland

As one can see, Serbia is one of, current six countries included in the Southeast European Hub that is established in 2019.

More information about organization and activities of this society can be found at the website https://www.europlanet-society.org/.

2. SERBIAN EUROPLANET GROUP

The Serbian Europlanet Group (SEG) currently consists of 20 members from 6 institutions. Details of members and activities of SEG can be found at https://www.europlanet-society.org/europlanet-society/regional-hubs/southeast-europe/.

The main activities of Serbian scientists in the Europlanet were:

- Organization of two Europlanet meetings and one session,
- Establishing of SEG webpage,
- Participations in the Europlanet science congresses and meetings,
- Participations in the Europlanet NA1 Expert Exchange Program, and
- Participations in the Europlanet commities.

In this paper we describe these activities and present scientific research of SEG members related to the Europlanet fields.

3. EUROPLANET MEETING ORGANIZATIONS

Serbian scientist organized two Europlanet workshops in Petica Science Center near Valjevo in Serbia:

- "Geology and geophysics of the solar system bodies" (24 June– 1 July, 2018), and
- "Integrations of satellite and ground-based observations and multidisciplinarity in research and prediction of different types of hazards in the Solar system" (10-13 May, 2019),

and Europlanet session during the XII Serbian-Bulgarian Astronomical Conference (XII SBAC) in Sokobanja, Serbia 25-29 September, 2020.

3.1. Europlanet workshops

3.1.1. Workshop in Geology and Geophysics of the Solar System

The workshop took place in Petnica Science Center, Petnica, Serbia (23 June – 1 July 2018) and further details can be found at http://petnica.rs/planetary2017. It was designed to cover a wide range of topics related to the formation, structure and dynamics of the Solar System and aimed to attract students and young researchers of various backgrounds and of different levels of experience in the fields of planetary sciences and space exploration. The workshop attended 43 participants, of which 24 PhD, 13 master and 6 undergraduate students. They were from 19 different home countries, including 15 from Eastern Europe, 3 from Russia and 4 from Northern Afrika. Other participants came from as far as India, Australia, and USA. The scientific organizers of the workshop were Dr. Katarina Miljkovic (Curtin University, Australia), Dr. Ana Cernok (The Open University, UK) and Dr. Matija Cuk (SETI Institute, USA), supported by the local organizers Dusan Pavlovic (Petnica Science Centre, Serbia) and Andrea Rajsic, deputy (University of Belgrade, Serbia). In total, there were 14 lecturers (7 female and 7 male). Although there was only one lecturer from a Serbian institution (University of Belgrade), there were 5 other lecturers (including the organizers) who were originally from Serbia. This planetary sciences workshop was supported by the Europlanet 2020 RI NA1 (Innovation through Science Networking) Task 5 (Coordination of ground-based observations) and Europlanet 2020 RI NA2 (Impact through outreach and engagement).

3.1.2. Workshop in Hazards in the Solar system

This workshop was focused on integrations of satellite and ground-based observations and multidisciplinarity in research and prediction of different types of hazards in the Solar system. The main of this meeting was connection of young researchers and scientists from under-represented countries, and experts in corresponding scientific fields. The organizer was the Geographical Institute "Jovan Cvijic" of Serbian Academy of Sciences and Arts. The chairs of the Scientific committee were Aleksandra Nina, Milan Radovanović from Serbia and Giovanni Nico from Italy. In this committee participated 11 scientists from 9 countries. Aleksandra Nina and Milan Radovanović were co-chairs, and Gorica Stanojević, Vladimir Čadež, Dejan Doljak, Vladimir Srećković and Dragoljub Štrbac were members of the Local Organizing Committee. The meeting attended 33 participants (of which 11 early career scientists) from 8 European countries: Bulgaria, Croatia, Greece, Hungary, Italy, Russia, Ukraine and Serbia. Their research fields relate to different theoretical and observation areas as well as to data sciences. In addition, two participants were from industry. This event was supported by the Europlanet 2020 RI NA1 - Innovation through Science Networking, Task 2: Scientific working groups (Europlanet 2020 RI has received funding from the European Union's Horizon 2020 research and innovation programme under grant No. 654208) and the Ministry for Education, Science and Technological Development of Republic of Serbia. More information about this event can be found at http://www.gi.sanu.ac.rs/site/index.php/en/activities/conferences-organisation/998-hazards-sos.

3.3. Europlanet session organised by SEG

Serbian scientists organized a Europlanet session during XII Serbian-Bulgarian Astronomical Conference (SBAS 12) that was held in Sokobanja from 25-29 of September 2020 (see Popović et al. 2020). Several lectures were held, a discussion, as well as the report of work of our group in the previous period was presented. At this Europlanet special session and during SBAC 12, possible directions for expanding cooperation were discussed with Bulgarian colleagues and also with colleagues from Europlanet Southeast HUB countries.

3.4. Participations in the Europlanet Science Congresses

Serbian scientists participated at the Europlanet Science Congresses (EPSC). The number of participants from Serbia is increasing. 7 scientists from Serbia participated in the EPSC-2020 and presented 3 lectures.

Figure 1: Participants of the workshop "Integrations of satellite and ground-based observations and multidisciplinarity in research and prediction of different types of hazards in the Solar system" held in Petnica Science Center on 10-13 May, 2019 (photo: Veljko Vujičić). From left to right:
Upper row: Konstantinos Kourtidis (Greece) , Pál Gábor Vizi (Hungary), Jelena Petrović (Serbia), Anđelka Kovačević (Serbia), Duško Borka (Serbia), Gorica Stanojević (Serbia), Zorica Marinković (Serbia), Bratislav Marinković (Serbia) and Dejan Doljak (Serbia);
Middle row: Georgi Simeonov (Bulgaria), Inna Pulinets (Russia), Bozhidar Srebrov (Bulgaria), Dejan Vinković (Croatia), Yaroslav Vyklyuk (Ukraine), Pier Francesco Biagi (Italy), Aleksandra Kolarski (Serbia), Lelica Popović (Serbia), Nikola Veselinović (Serbia) and Zoran Mijić (Serbia);
Bottom row: Predrag Jovanović (Serbia), Vesna Borka Jovanović (Serbia), Sergey Pulinets (Russia), Milan Radovanović (Serbia), Aleksandra Nina (Serbia), Vladimir Srećković (Serbia), Giovanni Nico (Italy), Milan S. Dimitrijević (Serbia), Luka Č. Popović (Serbia), Nataša Todorović (Serbia), Slavica Malinović-Milićević (Serbia) and Dragoljub Štrbac (Serbia).

3.5. Participations in the Regional Hubs Meetings

On the 4th and 5th June 2019, in Hotel Gellért, Budapest the Regional Hubs Meeting was organized by Melinda Dósa from Wigner RCP with the presence of the representatives from the Europlanet Society, Benelux Hub (represented by Ann Carine Vandaele, vice-president of Europlanet Society), Central European Hub, France, Italy, Northern European Hub, Spain & Portugal and Southeast European Hub, in total there were 23 researchers present. After the participants introduced themselves, the talk by Anita Heward, communication officer, was given on the Role of the hubs in the Europlanet Society and building a sustainable future from Europlanet 2020 RI. Following discussion was about the importance of widening.in Europlanet. The focus of the meeting was on Planetary science – technology – industry synergy: aims and possibilities & Towards a strategy definition. As a result of this meeting the participation and formal enrollment in Europlanet Society by Serbian researchers has been substantially increased.

3.6. Participation in the Europlanet NA1 Expert Exchange Program

Supported through the Europlanet NA1 Expert Exchange Program, Dr. Alena Zdravković, curator of the Mineral and Rock Collection of the Faculty of Mining and Geology in Belgrade, Serbia, visited The Open University in October 2017 to work with Dr. Ana Cernok and other experts in meteorite science. During this visit, six meteorite samples from the Marquis de Mauroy collection of the Mineral and Rock Collection (01. Lancon, 02. Bath, 03. Powder Mill Creek, 04.Morrisyown Hamblen, 05. Merceditas and 06. Hex River, with numbers representing a handing number at the Open University) were used for polished thin- and thick-section preparation at the Open University, Milton Keynes, UK. Lancon and Bath are fragmented chondrite meteorites, Powder Mill Creek and Morrisyown Hamblen are mesosiderites, and Merceditas and Hex River are iron meteorites. Since the meteorite samples belong to a very old collection, dating from 1899, due to inadequate equipment and unprecise preparation facilities in the laboratory of Faculty of Mining and Geology in Belgrade, those kind of samples were never used for utilizing cut and polishing preparations. This visit aimed at meteorite thin-section preparation was an important milestone for this Serbian collection. It was the first such opportunity to open and present the collection to an international scientific community. More importantly, those are the only thin sections of meteorite samples available at Belgrade University, and will therefore serve as precious teaching material for students educating, as well as for initiating meteorite research.

3.7. Participations in the Europlanet committees

As a member of the Southeastern European Hub Committee, Aleksandra Nina participated in two teleconferences and one meeting of the Selection committee for Europlanet funding.

3.8. Other activities

Lecture during XIX Serbian astronomical conference (19 SAC) held at the Serbian Academy of Sciences and Arts in Belgrade, from October 13 – 17, 2020 (see Kovačević et al. 2020).

4. SOME STUDIES OF SEG MEMBERS

SEG members are scientists in four research fields: astronomy, geophysics, physics and geography. Here we present a few research that are in Europlanet areas.

4.1. Astronomy

4.1.1. The functional relation between mean motion resonances and Yarkovsky force on small eccentricities

We examined asteroid's motion with orbital eccentricity in the range (0.1, 0.2) across the 2-body mean motion resonance (MMRs) with Jupiter due to the Yarkovsky effect. We calculated time delays d*tr* caused by the resonance on the mobility of an asteroid with the Yarkovsky drift speed. We derived a functional relation that accurately describes dependence between the average time lead/lag d*tr*, the strength of the resonance *SR*, and the semimajor axis drift speed d*a*/d*t* with asteroids' orbital eccentricities in the range (0.1, 0.2). We analysed average values of d*tr* using this functional relation comparing with obtained values of d*tr* from the numerical integrations. On the basis of the obtained results and analyses, we conclude that our equation can be used for the 2-body MMRs with strengths in the range $[1.3\times10^{-8}, 2.2\times10^{-4}]$, for Yarkovsky drift speeds in the range $[2.6\times10^{-4}, 2\times10^{-3}]$ au/Myr and for asteroids' orbital eccentricities in the range (0.1, 0.2) (Milić Žitnik 2020a).

4.1.2. The specific property of motion of resonant asteroids with very slow Yarkovsky drift speeds

We examined the specific characteristics of the motion of asteroids with very slow Yarkovsky drift speeds across the 2-body MMRs with Jupiter, whose strengths cover a wide range. It was found that the test asteroids with very small Yarkovsky drift speeds moved extremely rapidly across MMRs (order of magnitude 10^{-5} au/Myr or less). This result may indicate that, below a certain boundary value of d*a*/d*t* asteroids typically move quickly across MMRs. From the obtained results, it is concluded that the boundary value of the Yarkovsky drift speed is 7×10^{-5} au/Myr (Milić Žitnik 2019).

4.1.3. The relationship between the 'limiting' Yarkovsky drift speed and asteroid families' Yarkovsky *V*-shape

We examined the relationship between asteroid families' *V*-shapes and the 'limiting' diameters in the (a, $1/D$) plane. Following the recently defined 'limiting' value of the Yarkovsky drift speed at 7×10^{-5} au/Myr, we decided to investigate the relation between the asteroid family Yarkovsky *V*-shape and the 'limiting' Yarkovsky drift speed of asteroid's semi-major axes. We have used the known scaling formula to calculate the Yarkovsky drift speed in order to determine the inner and outer 'limiting' diameters (for the inner and outer *V*-shape borders) from the 'limiting' Yarkovsky drift speed. The method was applied to 11 asteroid families of different taxonomic classes, origin type and age, located throughout the Main Belt. Our main conclusion is that the 'breakpoints' in changing *V*-shape of the very old asteroid families, crossed by relatively strong MMRs on both sides very close to the parent body, are exactly the inverse of 'limiting' diameters in the a versus $1/D$ plane. This result uncovers a novel interesting property of asteroid families' Yarkovsky *V*-shapes (Milić Žitnik 2020b).

4.1.4. Improvement of modelling of atmospheres using A&M data

We continued to work on topics of modelling various atmospheres (using new software packages and supercomputers) and diagnostic of the astrophysical (terrestrial and space) and laboratory plasma using A&M datasets e.g. rate coefficients, Stark broadening parameters, line profiles (the shape of atomic spectral lines in plasmas contains information on the plasma parameters, and can be used as a diagnostic tool), etc. Results which are of interest for Europlanet community are presented in our recently published papers (see e.g. Ignjatović et al. 2019, Srećković et al. 2020, Majlinger et al. 2020, Dimitrijević et al. 2020) as well as in database MolD http://servo.aob.rs/mold (Marinković et al. 2019) hosted on SerVO at AOB.

4.1.5. Correlation of solar wind parameters with cosmic rays observed with ground station

It has been well known for more than half a century that solar activity has a strong influence on galactic cosmic ray (GCR) flux reaching Earth (anti-correlation). Coronal mass ejections (CMEs) structure and shockwave can additionally modulate GCRs, which could results in a transient decrease in observed GCR intensity, known as Forbush decrease (FD). These FDs can be detected even with ground muon detector (Savić et al. 2019). Variation of GCR can be analyzed correlating in situ measurement of the particles species present in solar wind with ground observations. Correlation between the 1-hour variations of GCR and several different one-hour averaged particle fluxes was found during FDs and it depends on energy of the particles of the solar wind as well as cut-off rigidities of secondary cosmic rays detectors on ground.

4.1.6. Habitability of exoplanets

Balbi, Hami and Kovačević (2020) present a new investigation of the habitability of the Milky Way bulge, that expands previous studies on the Galactic Habitable Zone. This work discusses existing knowledge on the abundance of planets in the bulge, metallicity and the possible frequency of rocky planets, orbital stability and encounters, and the possibility of planets around the central supermassive black hole. The paper focuses the two aspects that can present substantial differences with respect to the environment in the disk: (i) the ionizing radiation environment, due to the presence of the central black hole and to the highest rate of supernovae explosions and (ii) the efficiency of putative lithopanspermia mechanism for the diffusion of life between stellar systems. Authors devised analytical models of the star density in the bulge to provide estimates of the rate of catastrophic events and of the diffusion timescales for life over interstellar distances.

This article has been published as an invited contribution in the Special Issue "Frontiers of Astrobiology" edited by Manasvi Lingam.

Another concern for habitability is the presence of the supermassive black hole in the Galactic center, but also in nearby Active galactic nuclei, that could have resulted in a substantial flux of ionizing radiation during its past active phase, causing increased planetary atmospheric erosion and potentially harmful effects to surface life as shown by Wisłocka, Kovačević, Balbi (2019).

The goal of this paper is to improve our knowledge of the erosion of exoplanetary atmospheres through radiation from supermassive black holes (SMBHs) undergoing an active galactic nucleus (AGN) phase.

Authors extended the well-known energy-limited mass-loss model to include the case of radiation from AGNs. In the paper was calculated the possible atmospheric mass loss for 54 known exoplanets (of which 16 are hot Jupiters residing in the Galactic bulge and 38 are Earth-like planets, EPs) due to radiation from the Milky Way's (MW) central SMBH, Sagittarius A* (Sgr A*), and from a set of 107 220 AGNs generated using the 33 350 AGNs at $z < 0.5$ of the Sloan Digital Sky Survey database.

It was found that planets in the Galactic bulge might have lost up to several Earth atmospheres in mass during the AGN phase of Sgr A*, while the EPs are at a safe distance from Sgr A* (>7 kpc) and have not undergone any atmospheric erosion in their lifetimes. It was also found that the MW EPs might experience a mass loss up to 15 times the Mars atmosphere over a period of 50 Myr as the result of exposure to the cumulative extreme-UV flux FXUV from the AGNs up to z

= 0.5. This work was featured in famous *Forbes Magazine* in their section Innovation.

4.2. Geophysics

4.2.1. Investigation of a possible new type of lower ionosphere precursor of earthquakes

Analysis of the signal transmitted in Italy and received by the AbsPAL receiver in Belgrade in the period around the earthquake that occurred in the vicinity of Kraljevo on November 3, 2010 indicated a change in the amplitude of the signal less than an hour before this event. Although this change has not been reported in the literature, an additional study of several earthquakes indicates the existence of this change in other cases as well. The first study of this phenomenon is presented in Nina et al. (2020), and a broader statistical analysis is underway.

4.2.2. Modelling of solar X-ray flare influence on propagation of satellite signals

Due to the low electron density, the unperturbed D-region has practically no effect on the propagation of satellite signals. Therefore, it is generally not involved in modeling of signal propagation path or, if it is, its influence is given by analytical expressions based on observational data from higher altitudes. In Nina et al. (2020b), it is shown that during intense perturbations of this ionospheric layer due to the influence of solar X-ray flares (they do not perturb significantly higher ionospheric layers except when their intensity is very strong) it is necessary to include observational data for the D-region in modeling the propagation of satellite signals.

4.2.3. Satellite radar technique for atmospheric water vapor measurement and modelling effects of the ionospheric disturbances

Atmospheric water vapor measurement can be carried out in many different ways. One of the techniques for observing and measuring atmospheric water vapor is through satellite radars, precisely the Synthetic Aperture Radar (SAR) used and carried on the platform of many active satellites. In Radović (2020) are introduced four of such satellites and the water vapor modelling technique called SAR Interferometry is described as well. Along with the above mentioned in Radović (2020) it is demonstrated how neglecting the ionospheric disturbances that can occur during the satellite radar measurement of the water vapor can influence the modelling of certain parameters which are connected to the measured atmospheric water vapor.

4.2.4. Remote sensing of the atmospheric aerosol

Atmospheric aerosol plays one of the most important roles in climate changes and environmental issues through direct (scattering and absorption of solar and terrestrial radiation) and indirect (modification of cloud condensation nuclei through aerosol-cloud interaction) effects. In Mijić and Perišić (2019), study the relationship between satellite aerosol optical depth (AOD) measurements by Moderate Resolution Imaging Spectroradiometer (MODIS) and PM (Particulate Matter) concentrations data set from the Belgrade region was investigated. The preliminary results showed that AOD retrieved from a satellite sensor can be considered as a good proxy for ground observed PM mass concentrations. Within the EARLINET (European Aerosol Research Lidar Network) network a stand-alone lidar-based method (Papagiannopoulos et al. 2020) for detecting airborne hazards for aviation in near real time (NRT) is developed. In addition, Belgrade lidar station has been involved in ESA ADM-Aeolus mission (the first high-spectral resolution lidar in space) Cal/Val activity through validation of L2A products of aerosol and cloud profiles of backscatter, extinction and lidar-ratio.

4.2.5. Atmospheric disturbances due to severe stormy weather over Balkan region

Strong release of energy by atmospheric lightning discharges induced ionization changes along the propagation path of several Very Low Frequency (VLF) radio signals that had been received and recorded by Absolute Phase and Amplitude Logger (AbsPAL) system located in Belgrade (44.85^{o} N, 20.38^{o} E), at the Institute of Physics Belgrade, University of Belgrade, Serbia. Increased ionization is apparent in the perturbation of the signal amplitude and phase delay with respect to regular undisturbed ionospheric conditions. Integrated ground-based observations were performed with the aim to find coincidence and possible relationship between phenomena of VLF signal perturbations, optically documented Transient Luminous Events (TLEs) and documented lightning stroke events, during the stormy night of 27th-28th of May, 2009. The survey enclosed data from three independent sources: 1) VLF signal records from Belgrade Institute for Physics database, 2) video records of sprite events from ITALIAN METEOR and TLE NETWORK (I.M.T.N.) database and 3) detected lightning strokes from European Cooperation for Lightning Detection (EUCLID) network database. In most cases, the correspondence between VLF perturbations and CG strokes and on the other hand, VLF perturbations and TLE events, was found. In some cases the correspondence between all three phenomena was found (Kolarski 2019, 2020).

4.3. Physics

4.3.1. V. Čelebonović has been working on the problem of impact craters on the surfaces of solid planetary and satellite bodies. He showed that using standard solid state physics and measured properties of the craters, one can derive various parameters of the impactors. The calculations were checked on several known examples, and the agreement is reasonable.
4.3.2. The role of electron induced dissociation in the comet's coma and the findings during Rosetta spacecraft mission have been the subject of investigation published in Marinković et al. (2017). Data needs for modelling electron processes in cometary coma and their influence on the interpretation of the observed data by Rosetta instruments, have been discussed together with the currently available data and databases, where BEAMDB (Belgrade Electron/Atom(Molecule) DataBase - http://servo.aob.rs/emol) is given as an example (Marinković et al. 2019).

4.4. Geography

4.4.1. Our research was devoted to the determination of the causal relationship between the flow of particles that are coming from the Sun and the hurricanes Irma, Jose, and Katia. As a result of the preliminary analysis, using 12,274,264 linear models by parallel calculations, six of them were chosen as best. The identified lags were the basis for refinement of models with the artificial neural networks. Multilayer perceptrons with back propagation and recurrent LSTM have been chosen as commonly used artificial neural networks. Comparison of the accuracy of both linear and artificial neural networks results confirmed the adequacy of these models and made it possible to take into account the dynamics of the solar wind. Sensitivity analysis has shown that F10.7 has the greatest impact on the wind speed of the hurricanes. Despite low sensitivity of pressure to change the parameters of the solar wind, their strong fluctuations can cause a sharp decrease in pressure, and therefore the appearance of hurricanes (Vyklyuk, et al. 2019).

4.4.2. Forest fires that occurred in Portugal on 18 June 2017 caused several tens of human casualties. The cause of their emergence, as well as many others that occurred in Western Europe at the same time remained unknown. Taking into account consequences, including loss of human lives and endangerment of ecosystem sustainability, discovering of the forest fires causes is the very significant question. The heliocentric hypothesis has indirectly been tested, according to which charged particles are a possible cause of forest fires. We must point out that it was not possible to verify whether in this specific case the particles by reaching the ground and burning the plant mass create the initial phase of the formation of the flame. Therefore, we have tried to determine whether during the critical period, i.e. from 15–19 June there is a certain statistical connection between certain parameters of the solar wind and meteorological elements. Based on the 2 hourly values of the charged particles flow, a correlation analysis was performed with hourly values of individual meteorological elements including time lag at Monte Real station. The application of the Adaptive Neuro Fuzzy Inference System models has shown that there is a high degree of connection between the flow of protons and the analysed meteorological elements in Portugal. However, further verification of this hypothesis requires further laboratory testing (Radovanović et al. 2019).

5. CONCLUSION

In this paper we present activities of Serbian scientists in the Europlanet. We describe two Europlanet workshops organized in the Petnica Science Center: "Geology and geophysics of the solar system bodies" and “Integrations of satellite and ground-based observations and multi-disciplinarity in research and prediction of different types of hazards in Solar system” that occurred in 2018 and 2019, respectively, and the Europlanet session during XII Serbian-Bulgarian Astronomical Conference that occurred in Sokobanja 2020. In addition, we present other activities that were primarily aimed at connecting SEG members coming from six institutions as well as the promotion of the Europlanet and ESEEH organizations. Several studies relevant for the Europlanet research fields are presented in the second part of this paper.

Acknowledgments

This research is supported by the Europlanet. The authors acknowledge funding provided by the Institute of Physics Belgrade, Astronomical Observatory (the contract 451-03-68/2020-14/200002), Faculty of Mathematics University of Belgrade (the contract 451-03-68/2020-14/200104) through the grants by the Ministry of Education, Science, and Technological Development of the Republic of Serbia.

References

Balbi A., Hami M., Kovačević A.: 2020, The Habitability of the Galactic Bulge, *Life*, 10, 132-145.

Čelebonović V.: 2020, The origin of impact craters, some ideas. *Bulg. Astron. J*, 33, 21.

Drimitrijević M. S., Srećković V. A., Ignjatović Lj. M., Marinković B. P.: 2021, The role of some collisional processes in AGNs: rate coefficients needed for modeling, *New Astronomy*, 84, 101529.

Ignjatović Lj. M., Srećković V. A., Dimitrijević M. S.: 2019. The collisional atomic processes of Rydberg alkali atoms in geo-cosmical plasmas. *Monthly Notices of the Royal Astronomical Society*, 483(3), 4202-4209.

Kolarski A.: 2019, *Atmospheric disturbances due to severe stormy weather*, Book of Abstracts, “Integrations of satellite and ground-based observations and multi-disciplinarity in research and prediction of different types of hazards in Solar system”, Petnica, Science Center, May 10-13, 2019, Valjevo, Serbia, Geographical Institute “Jovan Cvijić” SASA, Belgrade, Serbia, p. 42.

Kolarski A.: 2020, *Storm activity over Balkan region during May 2009*, Book of Abstracts, "XII Serbian-Bulgarian Astronomical Conference (XII SBAC)" September 25-29, 2020, Sokobanja, Serbia, Astronomical Observatory, Belgrade, Serbia, p. 75.

Kovačević A., Kovačević Dojčinović A., Marčeta D., Onić D: 2020, Book of abstracts XIX Serbian Astronomical Conference October 13 - 17, 2020, Belgrade, Serbia.

Majlinger Z., Dimitrijević M. S., Srećković V. A.: 2020, Stark broadening of Co II spectral lines in hot stars and white dwarf spectra, *Monthly Notices of the Royal Astronomical Society*, 496(4), 5584-5590.

Marinković B. P., Bredehöft J. H., Vujčić V., Jevremović D., Mason N. J.: 2017, Rosetta Mission: Electron Scattering Cross Sections - Data Needs and Coverage in BEAMDB Database, *Atoms*, 5(4), 46 [18pp].

Marinković B. P., Srećković V. A., Vujčić V., Ivanović S., Uskoković N., Nešić M., Ignjatović Lj. M., Jevremović D., Dimitrijević M. S., Mason N. J.: 2019, BEAMDB and MOLD – Databases at the Serbian Virtual Observatory for collisional and radiative processes, *Atoms*, 7(1), 11 [14pp].

Mijić Z., Perišić M.: 2019, *Comparison of Modis aerosol observations and ground-based PM measurement for the Belgrade region*, Book of abstracts, "Integrations of satellite and ground-based observations and multi-disciplinarity in research and prediction of different types of hazards in Solar system", pp.51-52, Petnica Science Center, 10-13 May, 2019, Geographical Institute "Jovan Cvijić" SASA, Belgrade.

Milić Žitnik I.: 2019, The specific property of motion of resonant asteroids with very slow Yarkovsky drift speeds, *MNRAS*, 486, 2435-2439.

Milić Žitnik I.: 2020a, The functional relation between mean motion resonances and the Yarkovsky force for small eccentricities, *MNRAS*, 498(3), 4465-4471.

Milić Žitnik I.: 2020b, The relationship between the 'limiting' Yarkovsky drift speed and asteroid families' Yarkovsky V-shapes, *SAJ*, 200, 25-41.

Nina A., Radovanović M., Srećković V. A.: 2019, Book of abstracts, "Integrations of satellite and ground-based observations and multi-disciplinarity in research and prediction of different types of hazards in Solar system", Petnica Science Center, 10-13 May, 2019, Geographical Institute "Jovan Cvijić" SASA, Belgrade.

Nina A, Pulinets S., Biagi P. F., Nico G., Mitrović S. T., Radovanović M., Popović L. Č.: 2020a, *Sci. Total Environ.*, 710, 136406.

Nina A., Nico G., Odalović O., Čadež M., V., Todorović Drakul M., Radovanović M., Popović L. Č.: 2020b, *IEEE Geosci. Remote Sens. Lett.*, 17(7), 1198 – 1202.

Papagiannopoulos N., D'Amico G., Gialitaki A., Ajtai N., Alados-Arboledas L., Amodeo A., Amiridis V., Baars H., Balis D., Binietoglou I., Comerón A., Dionisi D., Falconieri A., Fréville P., Kampouri A., Mattis I., Mijić Z., Molero F., Papayannis A., Pappalardo G., Rodríguez-Gómez A., Solomos S., Mona L.: 2020, An EARLINET early warning system for atmospheric aerosol aviation hazards, *Atmos. Chem. Phys.*, 20, 10775–10789.

Popović L. Č., Srećković V. A., Dimitrijević M. S., Kovačević A.: 2020, *Book of Abstracts,* "XII Serbian-Bulgarian Astronomical Conference (XII SBAC)" September 25 - 29, 2020, Sokobanja, Serbia, Astronomical Observatory, Belgrade, Serbia

Radovanović M. M, Vyklyuk Y, Stevančević T, M, Milenković Đ. M, Jakovljević D, Petrović M, Malinović Milićević S, Vuković N, Vujko A, Yamashkin A, Sydor P, Vuković D, Škoda M.: 2019, Forest fires in Portugal — case study, 18 June 2017, *Thermal Science*, 23(1), 73-86.

Radović J.: 2020, *Satellite radar technique for atmospheric water vapor measurement and modelling effects of the ionospheric disturbances*, Master thesis, Faculty of Physics, University of Belgrade, Serbia.

Savić M., Veselinović N., Dragić A., Maletić D., Joković D., Banjanac R., Udovičić V.: 2019, Rigidity dependence of Forbush decreases in the energy region exceeding the sensitivity of neutron monitors, *Advances in Space Research*, 63(4), 1483-1489.

Srećković V. A., Dimitrijević M. S., Ignjatović Lj. M.: 2020, The influence of collisional-ionization and recombination processes on spectral line shapes in stellar atmospheres and in the hydrogen clouds in broad-line region of AGNs, *Contrib. Astron. Obs. Skalnaté Pleso*, 50, 171-178.

Vyklyuk Y, Radovanović M. M, Milovanović B, Milenković M, Petrović M, Doljak D, Malinovic Milićević S, Vuković N, Vujko A, Matsiuk N, Mukherjee S.: 2019, Space weather and hurricanes Irma, Jose and Katia, *Astrophys Space Sci,* 364, 154.

Wisłocka A. M., Kovačević A. B., Balbi A.: 2019, Comparative analysis of the influence of Sgr A* and nearby active galactic nuclei on the mass loss of known exoplanets, *Astron. Astrophys*., 624, A71 [17 pp].

Proceedings of the XII Serbian-Bulgarian Astronomical Conference
(XII SBAC) Sokobanja, Serbia, September 25-29, 2020
Editors: L. Č. Popović, V. A. Srećković, M. S. Dimitrijević and A. Kovačević
Publ. Astron. Soc. "Rudjer Bošković" No 20, 2020, 123-128

MILUTIN MILANKOVIĆ AND ASSOCIATES IN THE CREATION OF THE "*KANON*"

NATALIJA JANC[1], MILIVOJ B. GAVRILOV[2],
SLOBODAN B. MARKOVIĆ[2], VOJISLAVA PROTIĆ BENIŠEK[3],
LUKA Č. POPOVIĆ[3] and VLADIMIR BENIŠEK[3]

[1]*Baltimore, Maryland 21212, USA*
[2]*University of Novi Sad, Faculty of Sciences, Trg Dositeja Obradovića 3, 21000 Novi Sad, Serbia*
[3]*Astronomical Observatory, Volgina 7, P.O.Box 74, 11060 Belgrade, Serbia*
E-mail: natalijanc@earthlink.net

Abstract. Milutin Milanković took a different approach to climatology, when compared to other meteorologists of his time, and can be considered a key figure in laying down the foundations of modern climatology, where celestial mechanics was the foundation upon which Milanković based his theory of climate change.

Under the guidance of Vojislav Mišković, director of the Astronomical Observatory, mathematicians Stanimir Fempl and Dragoslav Mitrinović performed the lengthy and very comprehensive calculations. At the University of Belgrade in 1932 their scientific endeavor was completed. This work involved a multidisciplinary approach. Mathematician Mihailo Petrović Alas published a paper about this important project.

The problem of the shape of the Earth and the position of the Earth's poles was addressed by Milanković in 1932 and 1933, prompted by earlier suggestions of Alfred Wegener (1880–1930). Milutin Milanković published papers on the subject of *Mathematical Climatology* in significant scientific publications, such as the *Handbook of Climatology* and Gutenberg's *Handbook of Geophysics*. However, his works were not easily accessible to the interested scientists because few of the libraries had all the volumes of these Handbooks and other journals. The idea of creating his *Kanon* was presented at a meeting of the Academy of Natural Sciences in Belgrade in 1938. Mathematician Tatomir Andjelić did a tremendous amount of work during the preparation of Milanković's *Kanon*. *Kanon* was published in 1941 in Belgrade.

After World War II, under the heading *Open Problems*, Milanković presented 26 topics related to his work for further investigation to members of the Mathematical Institute, the Astronomical Institute, as well as graduate and doctoral students. Among them are several topics that are related to his *Kanon*.

1. INTRODUCTION

Milutin Milanković (1879–1958) opted for a different approach to climatology than meteorologists at the time. Therefore, it can be considered that he participated in laying the foundations of modern climatology. Celestial mechanics was the basis upon which he founded the theory of climate change (Milanković 1952).

A great deal of work regarding the scientific research of Milutin Milanković has been done at the Astronomical Observatory in Belgrade. Based on the work of the French astronomer Le Verrier (1811–1877), the calculations of secular changes in the astronomical elements of the Earth's trajectory were revised, taking into account the mass of each of the planets known until 1928.

2. ASSOCIATES IN THE CREATION OF THE "KANON"

Under the guidance of Vojislav Mišković (1892–1976), director of the Astronomical Observatory, mathematicians Stanimir Fempl (1903–1985), then an assistant, and Dragoslav Mitrinović (1908–1995), then a student, performed the lengthy and very comprehensive calculations using mechanical calculators (Janc et al. 2019) (Fig. 1). Their scientific endeavor was completed. It consisted of forming

Figure 1: The astronomer's desk at the Museum Collection located at the Meteorological Observatory. It features the Original Odhner Gothenburg, one of the desktop computing machines that was used to perform Milanković's calculations. (Foto: N. Janc, 1987.).

an approximate picture of the insolation of the Earth's surface, as well as the relationship that exists between the insolation and the temperature of both the Earth's surface and the atmosphere. The work involved mathematicians and astronomers who taught mathematical physics, celestial mechanics and astronomy.

Mathematician Mihailo Petrović Alas (1868–1943) published a paper *Occasion of a Recent Application of Astronomy to Climatology* (1932) on this important multidisciplinary project. He said:

"This year, a new scientific endeavor was completed at the University of Belgrade this year, with the cooperation of mathematicians and astronomers who teach mathematical physics, celestial mechanics and astronomy at the University's Faculty of Natural Sciences. The endeavor consisted in forming an approximate picture of the course of the insolation of the Earth's surface, as well as the relationship between the Sun's surface and the temperature of the Earth's surface on one side and the atmosphere on the other.

No matter how difficult this task may have been, having consisted in re-doing the work all over again, because the corrections made on the masses of the planets originated from their last calculations, and then recalculating values of secular inequalities for the elements of motion of the planet (which included 600,000 years before 1800), Mr. Milanković took it without hesitation. Assisted by Mr. V. V. Mišković, Director of the Astronomical Observatory of the University of Belgrade, who took over all astronomical calculations, could successfully complete this work on testing his new theory of climate change on Earth." (Petrović 1932).

Dragoslav Mitrinović, professor of mathematics at the Department of Electrical Engineering in Belgrade, wrote in his memories:

"For the purpose of scientific research, Professor M. Milanković undertook at the Astronomical Observatory in Belgrade, during 1928 and 1929, the tedious work of recomputing the secular changes of the astronomical elements of the Earth's trajectory, based on Le Verrier's work, and taking into account the values of the masses of planets known by 1928. The work was organized like this. Stanimir Fempl, then a university teaching assistant, and I, then a student, independently of one another, had to perform the proposed calculations." (Mitrinović 1968).

In his work *Milanković's Contribution to the Astronomical Theory of Ice Ages* (1979), Stanimir Fempl, professor of mathematics at the Faculty of Civil Engineering in Belgrade, writes that Milanković initially used Stockwell's results, but later used results of Mišković. "He (Mišković) used LeVerrier's calculations as more reliable, but he made corrections according to new knowledge about planetary masses. He did that with the cooperation of his assistants, Dragoslav Mitrinović and Stanimir Fempl. The calculations lasted for almost three years. Mišković also determined the degree of accuracy with which the calculations were performed" (Fempl 1979).

In 1930, Mišković initiated and edited the *Yearbook*, which, from the following year, was renamed the *Yearbook of Our Sky*. The same year the second edition of the *Annuaire de l'Observatoire astronomique de l'Université de Belgrade* was published. On this occasion, Milanković sent his congratulations to Mišković in a letter dated November 24, 1930. Milanković also wrote that he realized that some of the tables could be further elaborated, e.g., Table 5 *Length of day and night in the polar zones.* (*Lange Tage und Nächte der Polarzonen*) (Janc et al. 2018). He asked Mišković to prepare some more tables for him, which he marked on a separate piece of paper; unfortunately, that paper was not preserved (Janc et al. 2018). He needed the tables by the second half of December 1930 for his paper *The Earth Rotation* that he was preparing for the *Handbook of Geophysics* (Janc et al. 2018).

The problem of the shape of the Earth and the position of the Earth's poles began to be addressed by Milanković in 1932 and 1933, following the earlier suggestions of Alfred Wegener (Andjelić 1979). Milutin Milanković has published papers on the subject of *Mathematical Climatology* in significant scientific publications, among which are the *Handbook of Climatology* and the Gutenberg's *Handbook of Geophysics*. However, as he noted, his works were hardly accessible to interested readers because few of the libraries had all the volumes of these *Handbooks* and other relevant journals (Milanković 1952). So, he decided to publish all his papers on the paleoclimatic problem in a separate book (Milanković 1952). This is how the idea of creating the *Kanon* was born. The idea and content of the book were presented on March 27, 1938 at a meeting of the Academy of Natural Sciences in Belgrade, when a decision was made to publish it as an edition of the Serbian Royal Academy, in German, so that it would be accessible also to foreign scientists (Milanković 1952).

Tatomir Andjelić (1903–1993) was a professor of theoretical mechanics at Faculty of Science in Belgrade and academician of SANU. In the period 1928–1945 he worked as a high school professor of mathematics and at the same time he was a teaching assistant in rational mechanics at the University of Belgrade. He performed a lot of work in checking formulas, numerical tables, languages, etc. in *Kanon* (Trifunović 2007).

The *Borba* newspaper published on March 15, 1958 an article entitled *Palms and Bananas in Belgrade*, which claims that "In the next 100,000 years we will not reach the Ice Age – according to the mathematician Fempl" and quoted him as saying: "The results I have received are not only very interesting, but are encouraging as well. It turned out that in the future, in forty thousand years, the amount of heat emitted by the sun would constantly increase in our northern hemisphere. In the southern hemisphere, in the temperate zone, where New Zealand is located, the picture will be quite different. Very high minimums will appear, ten thousand and twenty thousand years from now."

3. PROBLEMS TO BE WORKED ON

Milutin Milanković left room for future associates on *Kanon*. Under the title *Problems to be worked on*, he presented 26 topics that could be addressed by members of the Mathematical Institute, the Astronomical Institute, as well as by graduate students and doctoral students (Trifunović 1979). Given that Milanković mentions the Mathematical Institute, the list of topics must have emerged after the year 1946 (Trifunović 1979). Some of the topics:

For the Mathematical Institute

A model of secular insolation on Earth

For the Astronomical Institute

A new study of secular perturbations for the past and the future

A new determination of the aberration constant using extragalactic objects

For graduate students

Insolation of the Earth's tropical zone, the caloric equator

Exactly calculating the secular course of insolation of the Earth over the past 50,000 years and as many future years

Insolation curve based on the theory from Chapter XVIII of the *Kanon*

For Ph.D. candidates

The exact calculation of the coefficient *m*

Calculation of the annual insolation of the parallels $\omega = \omega\ (t)$; see "Theorie mathématique." From here, heat parameters are calculated… from the equation $\omega\ (t\ \ ) = \omega\ (t\ \ + T / 2)$.

The theory of meteors passing through the Earth's atmosphere

Calculation of the coefficients b_0, b , b , b of Table VI ("Kanon", p. 312) using the method reported on p. 313–315 of "Kanon."

Study of atmospheric circulation, computational and by a model. To start with steady state. Mean annual temperatures (or insolation) on the parallels

Investigate separately the effect of changes in the inclination of the ecliptic on the insolation of the Earth with the assumption of a circular path, and then separately the influence of the change of eccentricity. That would be a quite simple derivation for geologists.

The problem of rolling snowballs

The problem of two bodies, §5 "Foundations of Celestial Mechanics" if M and therefore μ are considered to be variable (linear)

The list indicates that Milanković himself saw the need for certain topics in the *Kanon* to be processed in a modern way, given the new scientific data and numerical methods using modern computers (Trifunović, 1979).

4. CONCLUSION

The *Kanon* of Milutin Milanković was of crucial importance for understanding climate change and its causes. Under the leadership of Milanković, a multidisciplinary team was engaged in the realization of his idea and research, as described by Mihailo Petrović Alas (1932). People from several disciplines were involved in various phases, both in terms of professional education and academic title, from students to PhD scientists and academics.

This paper precisely emphasizes this aspect of cooperation, which, as Trifunović (2007) writes, was the first case of teamwork in Serbian science, which helped to complete the exceptional work of Milutin Milanković.

References

Andjelić T.: 1979, *Milutin Milanković biografija*, digitalni legat, Galerija SANU, 36.

Fempl S.: 1979, Milankovićev doprinos astronomskoj teoriji ledenih dobi, *Vasiona.*

Janc N., Protić-Benišek V., Benišek V., Gavrilov M. B., Popović L. Č., Marković S. B.: 2018, Academicians Milutin Milanković and Vojislav Mišković: Correspondence about Alfred Wegener and Wladimir Köppen, *Astronomical and Astrophysical Transactions*, 30(4), 505–510.

Janc N., Gavrilov M. B., Marković S. B., Protić-Benišek, Benišek V., Popović L. Č., Tomić N.: 2019, Ice age theory: correspondence between Milutin Milanković and Vojislav Mišković, *Open Geosciences*, 11(1), 263–272.

Milanković M.: 1952, *Uspomene, doživljaji i saznanja iz godina 1909 do 1944,* SAN, Posebna izdanja, Knjiga CXCV, Odeljenje prirodno-matematičkih nauka, Knjiga 6, Beograd.

Mitrinović D.: 1968, *Život Mihaila Petrovića,* Matematička biblioteka, Beograd, vol. 38.

Trifunović D.: 1979, Naučne teme koje je Milutin Milanković ostavio otvorenim, *Dijalektika,* XIV, 3–4.

Trifunović D.: 2007, *Iz prepiske Milutina Milankovića 1879–1958*, BeoSing, Beograd, 320.

Petrović M.: 1932, A propos d'une récente application de l'astronomie à la climatologie, *Publications de l'Observatoire astronomique de l'Université de Belgrade,* I, 7–22.

Pešić N.: 15-III-1958, Palme i banane u beogradskim parkovima…, *Borba*, Beograd.

Proceedings of the XII Serbian-Bulgarian Astronomical Conference
(XII SBAC) Sokobanja, Serbia, September 25-29, 2020
Editors: L. Č. Popović, V. A. Srećković, M. S. Dimitrijević and A. Kovačević
Publ. Astron. Soc. "Rudjer Bošković" No 20, 2020, 129-138

MYTHOLOGICAL ORIGIN OF CONSTELLATIONS AND THEIR DESCRIPTION: ARATUS, PSEUDO-ERATOSTHENES, HYGINUS

MILAN S. DIMITRIJEVIĆ[1] and ALEKSANDRA BAJIĆ[2]

[1]*Astronomical Observatory, Belgrade*
[2]*Society for archaeoastronomy, and ethnoastronomy, "Vlašići", Belgrade*
E-mail: mdimitrijevic@aob.rs, aleksandra.bajic@gmail.com

Abstract. Didactic poem, *Phaenomena* writen by Aratus of Soli, *Catasterismi* (Καταστερισμοί) the only surviving scripture associated before with Eratosthenes of Cyrene and *De Astronomica*, also known as *Poeticon Astronomicon,* attributed earlier to the Roman historian Gaius Julius Hyginus, and their translation to Serbian, have been considered.

1. INTRODUCTION

A long, didactic poem, *Phaenomena* writen by Aratus of Soli (Ἄρατος ὁ Σολεύς; *c.* 315/310 BC – 240 BC) is the oldest preserved astronomical text in Europe, created about 270 BC. The macedonian king, Antigonus II Gonatas (*c.* 319-239 BC) ordered and financed this work. Aratus sought and found the sources of astronomical knowledge in the work with the same name of Eudoxus of Cnidus, which he transformed in a poem, making it easier to read and remember. In the following times his poem became very popular, gladly read throughout ancient Greece and then Rome, often translated into Latin, which greatly increased the number of transcripts so that it has been preserved to these days, unlike the book of Eudoxus.

The similar description of mythical origin of constellations is *Catasterismi* (Καταστερισμοί) the only surviving scripture associated before with Eratosthenes of Cyrene (Ἐρατοσθένης ὁ Κυρηναῖος - *c.* 276 - *c.* 194 BC), the chief librarian at the Library of Alexandria, whose works were burnt down when it is burned and exist only in fragments. This text came to our time as an *epitome*, a short version of a larger work, and, the unknown author is named Pseudo-Eratosthenes. It is also a famous works of antiquity about heaven and, unlike the text of the similar content (*Phaenomena*) of Aratus, from which many mythological topics in this text have been taken, provides data on the number, and brightness of stars in the described constellations, so that represents a kind of the first preserved star catalogue of ancient Greece.

The third book with the similar content is *De Astronomica*, also known as *Poeticon Astronomicon,* attributed earlier to the Roman historian Gaius Iulius Hyginus, though the true authorship is disputed.

We translated in Serbian *Phaenomena, Catasterismi* and *Poeticon Astronomicon.* In this contribution we consider and discuss these three writings.

2. ARATUS OF SOLI AND HIS *PHAENOMENA*

Aratus is born in Cilicia (today in southeast of Turkey), in the Ionian colony of Soli, about 315 BC. He was a student of Menecrates from Ephesus and Philitas from Cos. He was in contact with the stoic philosopher Zeno from Athens, who probably recommended him to the Macedonian king Antigonus II Gonatas (c. 319-239) from Pella. He came to Pela around 276 BC, in the service of the king, as court's doctor. Antigonus most likely ordered and financed the famous work of Aratus, *Phaenomena*., created around 270 BC. Aratus visited also the court of the Antiochian king Antiochus I Soter of Seleucia (324/3 - 261 BC), where he spent some time, as well as Alexandria and other cultural centers of the eastern Mediterranean region. He passed away around 240 BC and has been burried in Soli. Recently, are discovered rests of his monumental grave.

Figure 1: Aratus of Soli, the marble sculpture in the Archaeological museum of Naples.

Aratus's didactic poem *Phaenomena* is based on the astronomical text with the same name of astronomer and mathematician Eudoxus from Cnidus (*c* 390 – *c* 337).

Eudoxus, a student of Plato, left us the first systematic description of the constellations, established the first sophisticated geometric model of the motion of celestial bodies, and significantly improved observational astronomy. He studied mathematics with Architas of Tarentum and after his stay in Asia Minor in Czicus he came to Athens, where he joined Plato's Academy. He later returned to his native Cnidus, where he built an observatory.

He tried to explain all the peculiarities in the motion of celestial bodies on the basis of a combination of uniform circular motions. According to the Eudoxus model, stars are located on a sphere, which once a day rotates around an axis passing through the Earth, and the movement of other celestial bodies is described by a combination of rotating spheres with each axis tilted relative to the previous one at a certain angle.

The Eudoxus system could not predict the movements of celestial bodies accurately enough. Also, he was not able to explain why the planets change their speed, as well as their brightness, since they are always at the same distance from the Earth, if their spheres are concentric. However, he advanced this science so much that the whole period up to Hipparchus is often called the period of Eudoxian astronomy.

Phaenomena is a long, didactic poem, with 1154 verses, divided in several chapters: Stars and constellations, Circles of celestial sphere, About planets, Risings and settings of stars and constellations. The last part refers to atmospheric phenomena and meteorological knowledge and beliefs. This is a book about astro-

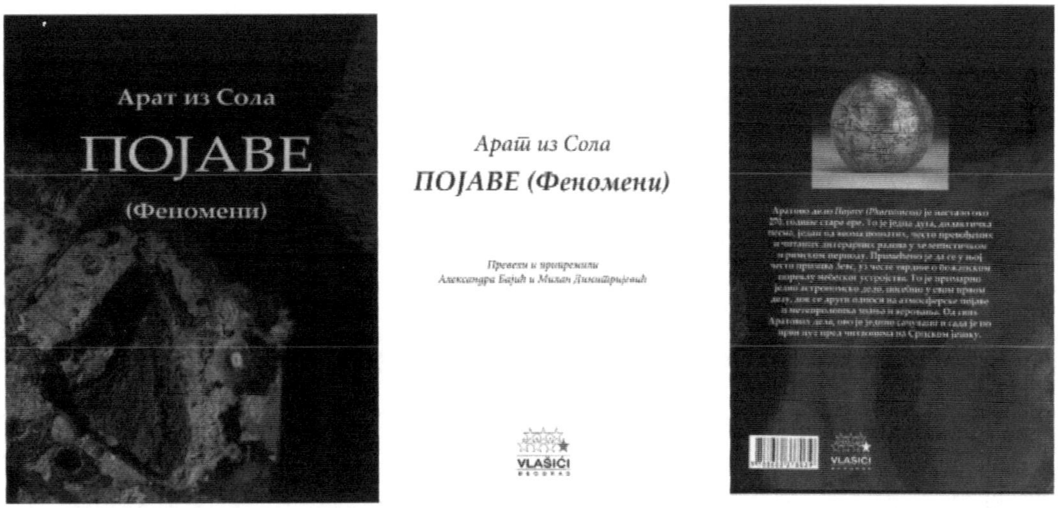

Figure 2: The fist translation of "Phaenomena" of Aratus of Soli on Serbian, published 2017.

nomical knowledge of Hellenistic Greece, in the time whe they were only descriptive, when astronomers could only descrie what they saw on the sky.

Aratus's work was very popular and often commented and translated into Latin in the Hellenistic and Roman periods, so that it has been preserved to this day, unlike the original work of Eudoxus, on which the author relied. The most famous of the commentators is Hipparchus, who lived about a hundred years after Aratus, the "father" of mathematical astronomy, whose only preserved work was just the commentary of Aratus' poem *Phaenomena*, to which he added a commentary of the work of Eudoxus from Cnidus. The oldest preserved copy in the Greek language appeared six centuries later by Theon from Alexandria (c. 335 – c. 405). Many translations have been made in Latin, the best known by Cicero, Germanicus and Avenius.

The work of Aratus and Eudoxus still inspires scientists and raises several questions:

1. What are the sources of Eudoxus' astronomical data?
2. How old are the astronomical data that Eudoxus possessed?
3. To what extent did he make observations himself?
4. Did Eudox really make a star globe?
4. From where did he observe the sky?
5. What instruments were available to him?
6. How accurate are his astronomical data?

The answers were sought, among others, by Isaac Newton (17th-18th century), and many scientists after him.

The research to answer some of these questions was conducted by Ovenden (1966), Roy (1984) and Zhytomirsky (1999) and (2003). They came to the conclusion that the latitude of that hypothetical observational place can be approximately determined on the basis of those stars of the southern celestial hemisphere that are not visible from there, that is, those that are not mentioned in the work. Using this premise, they determined that the observation site was located at a latitude of about 36° N (plus or minus 1.5°), which best suits the island of Crete, although it may be in the southernmost parts of Asia Minor. The scientists were also interested in the observation time. Knowing the precession of the Earth's axis, as well as the change in its inclination, they wondered at what point the arrangement of the stars in the sky was exactly as described by Aratus. Using different methods, all three came to very surprising conclusions: Ovenden found that the observation had to be done in 2600 BC (plus or minus 800 years), because some constellations had the positions stated by Aratus just then; Roy determined that it was in the year 2000 BC (plus or minus 200 years), while Zhytomirsky's calculations were very similar to the latter.

Criticizing their works, Schaefer (2002) pointed out several remarks: first, that the authors, when determining the latitude of the place, from which the results would possibly be obtained, did not take into account the refraction of light, so that the position of stars reported by Aratus are different. When this is taken into account, along with a few other astronomical parameters, Schaefer obtained the result that the southern constellations, described in Aratus's work, could be observed from a latitude of 31-33°, which corresponds, for example, to Phoenicia and not Crete. The same author pointed out another remark on the works of Ovenden, Roy and Zhytomirsky: none of these three authors published the calculations on which their conclusions are based.

In 2006, Rousseau and Dimitrakoudis used computer software to analyze Greek myths relating to the stars and constellations, as well as the geographical terms mentioned in them as the scene of the events described in each individual myth. They understood this as potential places of observation of the sky and determined that the data used by Aratus agree somewhere with the time when Eudoxus lived, but sometimes they reach the year 2000 BC. From their work, it follows that Eudoxus, in writing his work, on which Aratus later relied, used sources obtained from different places and at different times. Denis Duke (2008), a mathematician and statistician from the University of Florida, performed a statistical analysis of the Eudox data given by Aratus in his poem. His calculations narrowed the time span of sky observations to between 1150 and 300 BC, but they did little to the problem of determination of the location of the observations.

In the same year Elly Dekker (2008), investigating the conditions necessary to make a star globe, analyzes whether they were fulfilled in Eudoxus' time. The author points out that the descriptive tradition, to which both Eudoxus and Aratus belong, will be replaced by the mathematical one in the following centuries, which will be followed already by Hipparchus. This leads to a certain misunderstanding of Eudoxus's data, so, guided by certain conventions, which already existed at that time, she believes that Eudoxus must not have respected them, when they did not exist yet. Decker underlines a certain degree of standardization of the constellation, which Hipparchus knew and did not exist in the time of Eudoxus.

The same author emphasizes the importance of the fact that, when observing the sunrise on the days of solstices and equinoxes, the observer cannot directly see the constellation in which it is located, because it is shaded by light, but is indirectly oriented towards the constellation from the ecliptic,a briefly seen on the eastern horizon before sunrise (helical sunrise). In order to know exactly where the Sun is at the time of sunrise, the observer needs to have a precise instrument for measuring time and the knowledge that the celestial sphere rotates by 15° every hour. The clepsydra certainly does not allow for such precision. If we add to this the fact that the constellations from the ecliptic were not standardized (to an angular range of 30°), nor was there a convention that the center of the constellation Aries is at a declination of 0°, it becomes clear that Eudoxus could have made a star globe of his own observations, but he could not claim high precision. Eudoxus was not even aware of the precession of the Earth's axis, due to which this zero point moves slowly over time. In the absence of precise tools, it takes too much self-confidence to ignore older data and rely solely on the results of your observations.

Figure 3: The Kugel Globe (perhaps a part of an armillary sphere?) is dated around II century BC. It may be the earliest celestial globe to survive from Classical antiquity. It was acquired 1996 and held for some time in the Gallery J. Kugel Antiquaries in Paris (France). The Kugel Globe, which is now in a private collection, was reportedly found in the area of Lake Van, the largest lake in Turkey, located in the far east of the country.

The conclusion is that Eudoxus could have made a star globe, but it is not at the same time proof that he really did it. This is inferred from indirect data, found on two star maps, one discovered in a manuscript from the 11th century AD, known as Aberistvid NLW 735, the other from Monza, created in the 12th century (MS B 24/163), which show star maps, apparently copied from the globe, whose constellation drawings are quite similar to Eudoxus' data. This indicates that, at the time of writing, there could still be in Europe a specimen of a globe, made according to Eudoxus.

Previous research has shown that Eudoxus most likely observed the sky himself, and that, based on his observations as well as those of his predecessors and teachers, he most likely made the oldest star globe, which unfortunately has not been preserved, but some, a couple of centuries younger, made in his style, fortunately exist, such as the famous "Kugel globe", from the second century BC.

Aratus's work is, now for the first time, translated in the Serbian language (Арат из Сола, 2017).

We used the translation on English of G. R. Mair (Aratus of Soli, 1921) and Greek original.[1]

3. *CATASTERISMI* OF PSEUDO-ERATOSTHENES

Catasterismi are the only surviving scripture associated with Eratosthenes of Cyrene (Ἐρατοσθένης ὁ Κυρηναῖος - c. 276 - c. 194 BC), Greek mathematician, geographer, poet, astronomer and theoretician of music, the chief librarian at the Library of Alexandria, whose works were burnt down when it is burned and exist only in fragments. This text came to our time as an epitome, a short version of a more extensive work, from the end of the first century of our era. In the Middle Ages and during the renaissance, it was believed that it refers to the more comprehensive lost Eratosthenian work. Today, it is generally thought that this is not true, although some scientists defend the opposite view, so often the author is called Pseudo-Eratosthenes.

Catasterismi (Καταστερισμοί) in Greek means placing among the stars and denotes the transformation of a hero or object into a star or constellation. It is one of the most famous works of antiquity about heaven and describes the mythological origin of the constellations, the planets and the Milky Way, but, unlike the text of the similar content (*Phaenomena*) of Aratus of Soli, from which many mythological topics in this text have been taken, provides data on the number, and shine of stars in each of the described constellations, and in a way represents the first preserved star catalogue of ancient Greece. In addition, it also gives indications on the appearance of a person who personifies the constellation so that we can consider them in the way they are conceived in the Hellenistic world. In chapters 1 to 42, 43 constellations were considered, of the 48 (including Pleiades) described by Ptolemy. Chapters 43 and 44 speak of five planets and the Milky Way.

We made the first translation of *Catasterismi* (Димитријевић, Бајић, 2019) to the Serbian language with appropriate comments. We used translation of Kondos (1997) on English, and of Halma (1821) on French where also the text in Ancient Greek is given, as well as the text of Olivieri (Pseudo-Eratosthenes, 1897) in Ancient Greek created on the basis of five complete manuscripts and one partial (*Fragmenta Vaticana*), which used and Kondos. The great contribution of Kondos is his attempt to identify the stars mentioned by Pseudo-Eratosthenes. We included his identifications of stars in our translation.

---

[1] Aratus, Phaenomena,
http://www.perseus.tufts.edu/hopper/text?doc=Perseus%3Atext%3A2008.01.0483

4. *DE ASTRONOMICA* OF HYGINUS

De Astronomica or *Poeticon Astonomicon* contains information about the knowledge of celestial bodies and their apparent movements, as well as the ways in which that knowledge was incorporated into the understanding of the world and the religion of the ancient Romans and Greeks. It was written by a certain Hyginus, which is quite clear, because the author's signature exists at the very beginning, together with a dedication to a certain M. Fabius. Unfortunately, the author did not use the obligatory Roman trinomial, which defines the identity of a person more precisely. Therefore, it is not entirely clear whether this is the same Caius Iulius Hyginus, the former slave of Octavian August liberated by him and appointed to the post of the Head of the Palatine Library and a friend of the poet, Ovid. There are indications that this is not exactly that Hyginus, because he lived from c 64 BC to 17 AD, while the author of this book lists the constellations in a very similar order, which was used by Ptolemy in his *Almagest*, two hundred years later. Therefore, it is justifiably suspected that the author of this book lived in the second century of the new era, although it must be admitted that he did not quote Ptolemy as the source of his data in any single place. These dilemmas regarding the identity of the author of this book cannot be resolved, because Ptolemy himself could have used the order of describing the constellations of an earlier author, whose work has not been preserved, which would not diminish the significance of his *Almagest.*

Lippincott (2011) noted that Germanicus Iulius Caesar (24 May 15 BC – 10 October AD 19), who in 4 AD wrote a Latin version of Aratus's *Phainomena,* corrected a number of the astronomical mistakes, criticized later by Hipparchus. Lippincott, as an additional argument, underlines the fact that „Eratosthenes was the Keeper of the great Library of Alexandria" and Hyginus was on the same duty in the Palatine Library in Rome. This facilitated to him to cite in his work 44 Greek authors as counted by Bunte.[2] Lippincott (2011) says that among Greek authors Hyginus cites the work of Eratosthenes 21 times, „with ample evidence of additional, uncredited use elsewhere".

In his preface to Book I of *De Astronomica*, Hyginus says that he wants to give clearer explanations of the celestial sphere than Aratus, as well as to examine these issues more deeply. In the beginning of the Book I is the dedication to a certain 'M. Fabius', and an overview of the topics which the author wants to discuss, followed by detailed description of the celesial sphere and the corresponding circles.

---

[2] *Hygini Astronomica*, ed Bernh. Bunte, Leipzig, Weigel, 1875, pp. 3-6.

Figure 4: Two pages from the Erhard Ratdolt edition (1492, Venice) of the "De Astronomica" showing woodcuts of the constellations Cassiopeia and Andromeda. http://www.usno.navy.mil/library/rare/rare.html.

Book II gives myths connected with 42 constellations and discuss the mythologies associated with the five planets and the Milky Way. Book III presents desiption of each constellation, with indications of the shape and position of the figure. Additionally, the positions of the stars relative to the figure itself, are provided. Book IV provides the position of the constellations on each celestial circle, the unequal division of the night and day and the risings and settings of the constellations relative to the signs of the zodiac. Hyginus considers also the movements of the Sun and the Moon and the five planets. The end of the manuscript has been lost. Lippincott (2011) assumes that at the end was the consideration of the Metonic cycle, finding an indication for it in the Preface of Book I where Hyginus speaks about Meton and the accuracy of his observations of lunar and solar movements (*Hygini Astronomica,* 17-20, on p. 21) .

De Astronomica was included in the texts used for elementary learning of astronomy so it has preserved in a significant number of copies. It is translated to great world languages and now, we translated t on Serbian. We used one of Latin textes (*Hygini Astronomica*, e1875) and Russian (Гигин, 1997), English (Condos, 1997) and French (Hygin, 1983).

These three manuscripts, Aratus's *Phaenomena*, *Catasterismi* of Pseudo-Erathosthenes and *De Astonomica* of Hyginus, first time translated in Serbian, give a view on Eudoxan astronomy and the corresponding developments from Aratus in IV century BC to Hyginus during the reign of Octavian Augustus.

References

Aratus of Soli: 1921, *Phainomena,* (translated and edited by G.R. Mair), Classical texts library.

Арат из Сола: 2017, *Појаве (Феномени)*, превели и обрадили Александра Бајић и Милан С. Димитријевић, "Влашићи", Београд, 1-146.

Condos Theony (Translation and Commentary): 1997, *Star Myths of the Greeks and Romans*, A Sourcebook Containing *The Constellations* of Pseudo-Eratosthenes and the *Poetic Astronomy* of Hyginus, Phanes Press, Grand Rapids, Michigan USA.

Dekker Elly: 2008, Featuring the First Greek Celestial Globe, *Globe Studies*, 55/56, 133-152.

Димитријевић Милан С., Бајић Александра: 2019, *ПСЕУДО-ЕРАТОСТЕНОВИ "КАТАСТЕРИЗМИ"*, Антика некад и сад: Домети цивилизације и траг антике, Друштво за Античке студије Србије, Београд, 65-94.

Duke Denis: 2008, Statistical Dating of the Phenomena of Eudoxus, *DIO, The International Journal of Scientific History*, 15, 7-25.

ГИГИН: 1997, *АСТРОНОМИЯ*, перевод и комментарии А. И. Рубана. Санкт Петербург, Изд-во "Алетейя".

Halma, l'abbé: 1821, *Αρατου Σολεος Φαινομενα, Θεωνοσ Σχολια, Ερατοσθενους Καταστερισμοι, Λεοντιου Σφαιρα, et Germanici Caesaris Phaenomena*, Chez Merlin, Paris.

Hygini Astronomica, ed Bernh. Bunte: 1875, Leipzig, Weigel.

Hygin: 1983, *L'Astronomie* Texte établi et traduit par A. Le Boeuffle, Paris.

Kristen Lippincott:2011, *The textual tradition of the De Astronomia of Hyginus,* https://www. thesaxlproject.com/assets/Uploads/Hyginus-full-commentary-text-6-November- 2011.pdf

Ovenden, M. W.: 1966, The origin of constellations*, Philosophical journal* (Glasgow), 3(1), 1-18.

Pseudo-Eratosthenis: 1897, *Catasterismi,* recensuit Alexander Olivieri, Mytographi Graeci, Vol. III, Fasc. I, Lipsiae: In aedibus B. G. Teubneri

Rousseau, A., Dimitrakoudis, S.: 2006, A study of catasterisms in the Phaenomena of Aratus, *Medditeranean Archaeology and Archaeometry*, Special Issue, 6(3), 111-119.

Roy, A.: 1984, The Origin of the Constellations, *Vistas in Astronomy,* 27(2), 176-185.

Shaefer, B. E.: 2002, The latitude and epoch for the formation of the southern Greek constellations, *Journal for the History of Astronomy*, 33, Part 4, No. 113, . 313 – 350.

Zhytomirsky Sergey: 2003, *Phaenomena of Aratus, Orphism, and Ancient Astronomy,* In: Calendars, Symbols, and Orientations: Legacies of Astronomy and Culture, p. 79.

Zhytomirsky Sergey: 1999, Aratus' "Phaenomena": Dating and Analysing its Primary Source, *Astron. Astrophys. Transactions,* 17, 483-500.

**Proceedings of the XII Serbian-Bulgarian Astronomical Conference
(XII SBAC) Sokobanja, Serbia, September 25-29, 2020
Editors: L. Č. Popović, V. A. Srećković, M. S. Dimitrijević and A. Kovačević
Publ. Astron. Soc. "Rudjer Bošković" No 20, 2020, 139-153**

A PAIR OF STEĆAKS FROM DONJA ZGOŠĆA

ALEKSANDRA BAJIĆ[1] and MILAN S. DIMITRIJEVIĆ[2]

[1]*Society for archaeoastronomy, and ethnoastronomy, "Vlašići", Belgrade*
[2]*Astronomical Observatory, Belgrade*
E-mail: aleksandra.bajic@gmail.com, mdimitrijevic@aob.rs

Abstract. The study tests the hypothesis that artistic representations on some stećak tombstones can illustrate certain knowledge and beliefs from pre-Christian mythology, cosmology and religion, as well as the calendar knowledge of the Balkan Slavs. Stećaks are tombstones made of large marble monoliths. These are dated to the Middle Ages. Older researchers attribute them to the Bogumils, a religious sect whose religious beliefs are not sufficiently known, but that assumption has been abandoned. Now, it is believed that some of these were built by Christians (both catholics and orthodox), some by Muslims. Some are richly decorated with relief, some have inscriptions, written in Bosnian Cyrillic, in a language that is indisputably Slavic. Two stećak tombstones, found in the local cemetery in Donja Zgošća near Kakanj, in present-day Bosnia, are analyzed in this paper. These two are certainly not decorated with either Christian or Muslim symbols.

1. INTRODUCTION

Donja Zgošća is located in central Bosnia, north of Kakanj, on the banks of the river Zgošća. Two stećak tombstones, investigated in this paper, were found there in the old cemetery known as Crkvine, at geographical coordinates 44° 09 '17.5 "N, 18° 07' 54.5 E. The cemetery is heavily vandalized now, but, fortunately, these two stećak tombstones were transferred to State Museum in Sarajevo a long time ago. Both of these are richly decorated with relief.

2. GREAT STEĆAK

The stećak is made of marble monolith, large in size: 265 cm long, 139-147 cm wide, 146 cm high at the corners and 169 cm at the ridge. It stands on a slab about 40 cm thick, which was buried to ground level. It differs from all other stećak tombstones in the fineness, precision and richness of its construction. It is oriented east-west (Truhelka, 1933, p. 12), All its surfaces, except the upper one, are decorated with relief. Because of its beauty and artistic achievement, it is considered to be a monument to either a ruler or a high priest. There is no consensus among scientists about the time when it was made and it is not known who was buried under it. Radioisotope dating of the mortal remains of the deceased has not been done.

Figure 1: Western side (anti-facade) of the Great Stećak.

At the western side[1], we can see twelve rosettes, arranged in four verticals, consisting of three rosettes each. This can be understood as four seasons of three months each, which make up a solar year, therefore, as a symbolic representation of the calendar. That is why we should look at the facade, facing east:

The façade of the large stećak tombstone from Zgošća seems to illustrate the verses of a folk poem from the collection of Vuk Karadžić (*Women's Folk Songs*, Book 1, Poems Nos. 81 and 78):

O, Durmitor, o high mountain!
A white city was seen above you,
And above the city THREE SUNS were shining...

Vuk Karadžić (*Women's Folk Poems*, Book 1, Poem No. 78)

[1] Most of the photographs of the stećak tombstones were taken from the book by Alojz Benac, and Oto Bihalji-Merin (1964), when these monuments were in a much better condition, before the acid rains and before the experiments of the conservators.

Figure 2: The eastern side (façade) of the Great Stećak.

This is exactly what we see on the facade of the stećak: a city above a palisade of logs with a gate. Above the middle houses of the city, three rosettes are clearly visible, which can be understood as three Suns.

There was a belief among Serbs that three Suns were shining in the sky once. Nenad Đ. Janković wrote about it (1951, p. 38) referring to Vuk Karadžić and Simo Trojanović. Belief in the existence of three Suns, according to the same author, has its basis in the atmospheric appearance of parhelia. Christian books do not mention the trinity of the Sun, so this visual representation can only be understood as a remnant of the old religion of Balkan people.

In the collection of Vuk Karadžić, there is another variety of the poem that mentions those three Suns, which begins as follows:

O, Vitor, o high mountain,
Above you, THREE SUNS were rising ...

Vuk Karadžić (*Women's Folk Poems*, Book 1, Poem No. 81)

Mount Vitor does not exist in the Balkans. However, Christian missionaries who baptized the Slavs of Polab and Pomerania in the twelfth century AD mention the hill of Vithora, which was located above the city of Arkona, on the island of Rujan (today's Rügen). In the town of Arkona there was a temple of the four-headed Slavic God Svetovid (Swantevit), known on Balkans as Vid. The same missionaries also gave a description of the statue of Svetovid (Vid), which had four heads. That gave us the idea to look for Svetovid on this facade as well:

Three human figures are standing at the city gate and the fourth human figure can be seen hidden, in the first house on the left in the city. (The image is not sharp enough, but the right hand one of the three figures on the gate seems to hold in its right hand some round, hollow shape, resembling a ring.)

Figure 3: Ring in the hand of the first character on the right.

The representation of the ring is not a rarity on Herzegovinian stećak tombstones, on the contrary, we see it quite often, especially on those from Radimlja near Stolac, and it can also be seen in the hand of the famous idol from Zbruč, which is one of the few preserved visual representations of Svetovid. There are whole groups of such stećak tombstones, which depict a hero with a powerful, large either right or left hand, over which is a ring. The ring is obviously important. Why?

Figure 4: A detail of the stećak from Radimlja.

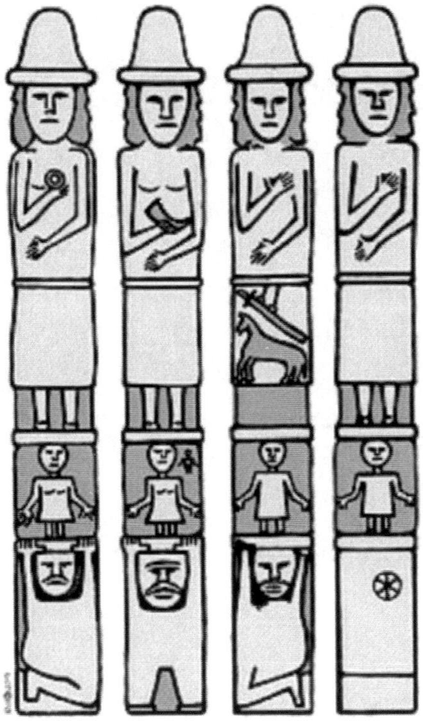

Figure 5: Four sides of the idol of Svetovid from Zbruch: one of his aspects has a ring in his hand.

The palisade below the city is very interesting: at first glance, it looks like a mountain, but that's not all. To the left of the gate, it consists of six logs (or posts), which, going towards the gate, grow in length. There are as many of these on the right side of the gate, a total of twelve, but starting from the gate to the right, their height is getting smaller. These could be associated with the twelve months in a solar year. If the gate represents the summer solstice, then the length of the logs illustrates well the length of the solar day in certain months. This reinforces the assumption of the calendar significance of the stećak tombstone. Therefore, we should look at the other elements on the facade:

On the "lower floor" of the façade, two horses can be seen, with human figures between them, but the stećak is damaged there, so it is not possible to determine how many of these figures there are. The left horse is without a rider, while on the right one a small human figure seems to be riding on the horse back and it looks like a child. (If there were three characters in the middle, then the total number is again four, if there were only two characters in the middle, then, with a small rider on the right horse, there are three in total.)

Of course, we should look at the north side of the stećak, in order to check this interpretation:

Figure 6: North side of the large stećak tombstone.

We will first look at the upper half of the northern side of the stećak, which is separated from the lower one by a border frieze. We can see five horsemen. Each of them has a hat:

The first of them is a bit bigger than the others, so it is probably more significant. It was marked with a plaque, on which there was an inscription once, which is known to have been intentionally destroyed (carved out) in the seventeenth century. The face (or faces) of that rider was also destroyed, which could mean that it was especially important.

Following the leading one, we can see three horsemen, each one with a rosette above. They are marked with a common plate from which the inscription is also destroyed. The second one of those three riders was damaged, only a part of his left leg along with buttocks and his sword can be seen, as well as a part of his right arm with a spear. Rosettes above the riders can signify three Suns.

The last rider is not marked by a nameplate, but his identity is revealed by the iris flower which is above him. It could be Perun, the Slavic God of Thunder. The Serbian word for iris, Perunika, is derived from Perun's name.

Thus, we get a "formula" that could interpret the four-faced Svetovid, who in the Balkans was called Vid.

Vid (quaternity) = Triglav (Trinity of the Sun) + Perun (the God of Thunder)

According to what we know about Svetovid (Vid) so far, he was the God of light, Sun and war. Saxon Gramaticus informed us in detail about this, in his work "Gesta Danorum." The same author stated that the festival of Svetovid was celebrated in the temple in Arkona "just before the harvest." It means that it was the time of the greatest power of the Sun. If we wanted to determine more precisely the day of the greatest power of the sun deity, it would be the summer solstice, when the day is the longest, and it is "just before the harvest."

On the lower half of the stećak's north side three scenes are depicted: deer hunting[2]; wild boar[3] hunting; a wolf, chained to a tree. The third one reflects the popular belief that in the World of the Dead there is a large tree (oak or hawthorn) to which a wolf is chained. The beast gnaws that chain, threatening to free itself. The moment the chain is almost broken, Christmas comes, someone shouts "Christ is born" and the chain is good as new again. There is also a belief that this happens when a gypsy blacksmith strikes his anvil with a hammer (Čajkanović, V, 1924, p. 132). Since the wolf, according to Carl Gustav Jung[4] (1943, p. 349), is a symbol of death and rebirth, the last scene almost certainly shows the world of the dead, which is confirmed by the above-mentioned folklore. Since Christmas is mentioned in that story, we realized that the scene also refers to the winter period. A dragon is above the wolf.

Figure 7: Stećak tombstones at the original location, the old cemetery in Zgošća.

---

[2] Deer hunting is a very common motif on stećak tombstones (Miletić, N: (1982), p. 74) and a favorite topic in mythology. Deer are often hunted in Serbian folk poem, most often in ritual wedding ones. King Arthur, from the Welsh myth (*The Story of Geraint, son of Erbin*) hunted a white deer on a holiday called Sul Gwyn, (meaning "white Sun" or "holy Sun"), which is the Welsh name of the summer solstice. Deer hunting was also found on Thracian frescoes. The deer symbolizes the Sun of the first half of the year, youth, the ruler himself and ritual purity. After sacrificing a deer, the ruler acquires the right to marry. In the calendar, the Sun ceases to be "young" from the summer solstice, the hot days of summer go by. The ruler is mature and in his full strength.

[3] Wild boar hunting is known by the myths of many peoples. Hercules hunted the wild boar, it is one of his twelve feats; the Vedic Indra hunted the wild boar, Emuza; King Arthur hunted the wild boar Twrch Trwyth (in The Story of Kilhwch and Olwen). South-Slavic folklore has not preserved a single story about a powerful hero who hunted a wild boar, but it is known that on Christmas even today it is obligatory to eat pechenitsa, ie pork roast, which may indicate a ritual sacrifice of a boar for the winter solstice.

[4] Karl Gustav Jung, (1943) *Psihologija i alkemija*, Naprijed, Zagreb, II izdanje, 1983. (C.G. Jung, Psychology and Alchemy).

The trees in this part of the stećak can be identified: deer hunting takes place next to a pine tree, the coastal one, with a wide canopy in the form of an umbrella. The wild boar hunt takes place near the tree from which the leaves are missing. The wolf is tied to an oak, the leaves of which have a quite recognizable shape.

Given that the figure of a deer hunter is just below the figure we thought to represent Svetovid, whose holiday was celebrated just in time of the summer solstice[5], we thought that a gnomon placed in the right place would be missing, whose shadow would "fall" on the first rider and deer hunter on the day of the summer solstice, at sunrise. Thus, this stećak tombstone could also have its practical use for determining the calendar. The assumption of the existence of a gnomon was confirmed: in the immediate vicinity of the large stećak tombstone from Zgošća, at the original location, there was another stećak tombstone, known as the Pillar from Zgošća, which is suitable to be a gnomon because of its shape. That stećak is also in the State Museum in Sarajevo.

This photograph is one of several preserved, showing both Zgošća stećak tombstones, in the original location.

3. THE PILLAR FROM DONJA ZGOŠĆA

This monument is also made of marble monolith, large in size. Its total height of about 300 cm. Its upper part is in the form of a four sided prism, Its cross section is rectangular, 54 x 44 cm, and its height 100 cm. The lower part of the pillar has a height of 200 cm, of which about 60 cm is buried in the ground. We can say that because of the absence of the decorative elements. The edges are cut at an angle of 45°, so its cross section is a slightly flattened octagon. The total height of the part above the ground is about 240cm, together with the transition zone between the upper and lower part, of which the lower, narrower part is about 140 cm. We notice that this height is approximately equal to the height of the side edge of the large stećak tombstone (146 cm).

The following photo shows the Pillar from Donja Zgošća, which is now also in the garden of the State Museum in Sarajevo, as well as the Great stećak tombstone. The term "original location" is only conditionally correct, because both these stećak tombstones were obviously moved there, and the large one was seriously damaged (vandalized) by somebody who was looking for buried treasure a long time ago.

---

[5] Saxon Gramaticus, who in his work "Gesta Danorum" gave a precise description of the temple of Svetovid, near the town of Arkona, on the island of Rujan (today's Rügen), literally says that the festival of Svetovid was celebrated "just before harvest." The same author states that Charlemagne burned the Temple of Svetovid in the 8th century for the first time and imposed on the Rujans the Christian cult of Saint Vitus (from Sicily), whose holiday was fixed for June 15 instead of the one of the Slavic God. At that time, the Julian calendar was exactly six days behind the natural, solar calendar, which would mean that the holiday was exactly on the summer solstice day.

Figure 8: The Pillar from Donja Zgošća.

The pillar-shaped stećak is sloping (see the figure no. 7), probably on the side where somebody was digging, probably with the same goal. At the same time, a part of it ornaments was buried in the ground, which means that it "sunk" more than it should.

Nevertheless, the joint photograph from the original location shows us that the two stećak tombstones are in some spacial relationship. The north side of the large stećak tombstone (the one with five riders and a deer hunter) faces the pillar-shaped stećak. According to the position of the shadows, we would say that the large stećak was oriented east-west. The same was stated by Dr Ćiro Truhelka, the former director of the State Museum, in his description of these stećak tombstones.

It is possible to place the PILLAR so that its summer solstice shadow covers the first rider, his nameplate and a deer hunting scene in the moment of sunrise. It should be located at an azimuth of about 56-57° in respect to the northeast corner of the large stećak, at any distance. When the first ray of the Sun appeared on the summer solstice, the shadow of the pillar would cover the first rider, his name-plate, as well as the entire scene of deer hunting. This all applies to the case that the horizon is plane (0°), about which there is no precise data, because no astro-geodetic survey was performed on the site. According to available photographs, the northeastern horizon is approximately plane.

If the pillar was positioned at the azimuth of 56-57° in respect to the front edge of the large stećak, we can see that the matching is complete. The shadow of the pillar will, at sunrise on the summer solstice, cover the entire scene of deer hunting and, above it, the first rider, the one who was marked with a special plate. (We have already assumed that the plaque read "Svetovid.") At the same time, if the pillar is placed at the appropriate angle, its distance from the large stećak is completely irrelevant.

This solution corresponds very well to the position seen in the photo from the original location.

There is another detail on the Pillar, which speaks in favor of the calendar significance of these monuments:

Figure 9: A detail from the Pillar.

These two lines on the pillar intersect at an angle of 67° (2x33.5°), which is the exact angle of the total shift of the sunrise point from the summer solstice to the winter solstice, at this geographical latitude.

All these calendar symbols on this pair of stećak tombstones are hardly a coincidence. The pattern on the Pillar shows four small rosettes (a qaternity) together with a unifying element (a larger rosette with a four-pointed star). Therefore, we should look at the other (southern) side of the large stećak tombstone from Zgošća to check all this (See figure 10).

The symbolism is similar to the symbolism seen on the north side, only this time we need to look from right to left. Again we can see five horsemen in a row, by the bottom of the stećak.

The first one is obviously different from all the others, he rides in the opposite direction from them and has a "Tree of the World" above him. This horsman is exactly opposite to the one we marked (on the north side) as Svetovid. Everything fits again, the Tree of the World is always tied to the supreme deity[6]. This horsman is directed from east to west, as befits a solar deity.

[6] An oak tree is dedicated to Zeus, an ash tree called Igdrasil to the Nordic Odin, and an oak tree to Svetovid.

Figure 10: The south side of the Great Stećak tumbstone.

The next three horsmen have, in the corner of the square in which they are located, one small rosette each in the shape of the Sun and two more in a rectangle above, in total, THREE each (three months of one season).

The last rider in the lower left corner has an iris flower in a square, and the flower wreaths are in a rectangle above him, arranged in two rosettes (not "Sun Wheels," but rosettes composed of flowers). Again, he could be identified as Perun and is located right behind the rider we marked as Perun on the opposite side of the stećak. This side of the stećak contains the same "formula" as on the one we commented on earlier:

Perun (thunder God) + Triglav (Trinity of the Sun) = Svetovid (quaternity)

All the symbolism of solar numbers exists on the large stećak thumbstone from Donja Zgošća:

12 - as twelve months of the solar year
3 - as THREE SUNS and TRIGLAV and three months in a season
4 - as four seasons and the quaternity of SVETOVID (VID)

The symbolism of visual representations also corresponds to the solar cult.

There is also the Tree of the World, so important in solar religious systems; There is also the iris plant, as Perun's sign; There is deer hunting (summer solstice); There is also wild boar hunting; There is a WOLF, a symbol of death, chained to the TREE OF THE WORLD – AN OAK; There is a DRAGON, a symbol of FIRE.

All this defines these two stećak tombstones as a unique calendar monument, closely connected with the observation of the sky and the determination of the cardinal days of the year (solstices and equinoxes).

Neither of these two stećak tombstones, neither the large one nor the one in the shape of a pillar, present any convincing Christian symbol, as well as Muslim ones: no cross, no crescent moon. The symbols, carved on the surface of these two monolith marble blocks can be "read" by following the old religion[7] of the Balkan Slavs, its iconography and mythology.

Looking carefully at the large stećak tombstone from Zgošća, we did not find any traces of the lunar calendar. If it is a calendar, then it is purely solar.

4. PARALLELS

There is only one stećak, very similar to the Great stećak. The construction is the same and the individual elements of its relief are closely resembling, both stylistically and symbolically, with those on the Great Stećak. Therefore, there is a high probability that they were carved by the same master, at approximately the same time. This stećak tombstone has not been published, so there is no available information about its original position. Now, it is located in front of the Kakanj Municipality building.

Figure 11: Stećak tombstone from the vicinity of Kakanj.

[7] We were NEVER claiming that all the stećak tombstones are pagan. We know that there are some of these with clear Christian symbols, as well as with clear Muslim symbols. Christianization of the inhabitants of Bosnia and Herzegovina was completed only by the end of the tenth century, when the Narentians (Neretljani in Serbian) were baptized by force, after they had been defeated in a war with Venice. It seems logical that remnants of the old religion can be expected a couple of centuries later, especially at remote places.

Three horsemen can be clearly seen, corresponding to those three on the stećak from Zgošća marked with a rosette, i.e. the trinity of the Sun. Above them, there is a border freeze of 12 three-leaf clovers, which is associated with 12 months of the solar year. The opposite side of the stećak should also be seen:

Figure 12: The human figure feeds the dragon "out of hand".

You can see a human figure, feeding a winged snake (dragon) out of his hand. This is probably Perun (the master of celestial fire, symbolically represented by the dragon). Again, he could be the fourth member of the quaternity, an aspect of the four-faced Svetovid. The facade of the stećak looks like this:

Figure 13: Facade of the stećak tombstone - unifying figure with a bow, an arrow and a ring.

We will look at another stećak, in the shape of a horizontal slab.

Figure 14: Stećak tombstone from Lištica.

Slab-shaped stećak tombstones are considered to be the oldest (Wenzel, Mariana 1962, 102-143). This one, shown in the previous picture, was found in Lištica, in Western Bosnia.

The crescent Moon is clearly visible, which is why it can be thought that celestial bodies are shown on it. There are also three larger rosettes, which can be understood as three Suns, just like on the Great Stećak from Zgošća. There are also three swastikas. Swastika is an ancient symbol of the sky, which would support the previous assumption. There is also a unifying symbol, a three-leaf clover, which can signify the unity of the three Suns (like Triglav). Next to the Moon's crescent is probably Venus, in the shape of the morning "star" Danica, or the evening, Večernjača, shown as a smaller rosette. It is astronomically true that Morning star (Danica) can be near the waning crescent only, while Evening star can only be near the waxing crescent Moon. It is always close to the Sun, at a maximum elongation of 47°.

Christianity was not very interested in the Moon (except when calculating the date of Easter). It was not interested in Venus either and certainly did not understand the Sun as triple. The visual representations on this stećak are a part of a different cosmology.

CONCLUSIONS

Sources for the study of the pre-Christian mythology of the South Slavs obviously exist, both in folk literature and in monuments of material culture. Within these sources, one can search for (and find) both knowledge and beliefs about the sky, celestial bodies and the structure of the world, as well as the calendar knowledge. Vuk Karadžić, in the second edition of his Dictionary (1852, under S), states that the people in the vicinity of Imotski call the stećak tombstones "Old Feith Stones" ("Kamenje Starovirsko"), which can also point to the old, pre-

Christian faith of the Balkan Slavs. At the creation time of the stećak tombstones, the autochthonous Romanized Illyrian population had been assimilated long since. There are no traces of the Illyrian language on these monuments. There is no Latin either.

An interpretation is good if it explains a large number of phenomena. Unfortunately, no one has researched either the symbols of celestial bodies or traces of pre-Christian cosmology and calendar knowledge on stećak tombstones so far. Therefore, it is not possible to cite multiple references for interpretations given in this paper.

Since being placed under the protection of UNESCO, stećak tombstones have become a "sensitive political issue" between Croats, Serbs and Bosniaks. Issues of origin and identity are never resolved without emotions, often exaggerated.

References

Bihalji Merin, Oto; Benac Alojz: 1964, *Steine der Bogumilen*, A. Seemannn-Verlag, Leipzig.

Čajkanović Veselin: 1924, *O vrhovnom Bogu u staroj srpskoj religiji*, SKZ, BIGZ, Prosveta, Partenon MAM, izdanje 1994, knjiga 1.

Gramaticus Saxon: 2009, *Gesta Danorum* (Prevod na Srpski jezik poglavlja, koja se odnose na Slovene: Rastko Kostić, *Pad Arkone i sumrak slovenskog paganizma,* 2009. Atos, Beograd)

Janković Nenad, Đ: 1951, *Astronomija u predanjima, umotvorinama i običajima Srba*, Odeljenje za društvene nauke SANU, Beograd.

Jung Karl, G: 1942, *Psihologija i alkemija*, Naprijed, Zagreb (II izdanje, 1984)

Karadžić Vuk St.: 1841, *Srpske narodne pjesme*, reprint Nolit, Beograd, 1989 (Knjiga 1, *Ženske narodne pjesme*).

Karadžić Vuk St: 1852, *Srpski rječnik,* II izdanje (reprint, Nolit, Beograd, 1977).

Miletić Nada: 1982, *Stećci*, Izdavački zavod Jugoslavije (Beograd) i Spektar (Zagreb) i Prva književna komuna (Mostar), edicija *Umetnost na tlu Jugoslavije.*

Truhelka Ćiro: 1933, *Kolijevka i groblje prvih Kotromanića*, Nastavni Vjesnik, XLI.

Wenzel Mariana: 1962, *Bosnian and Herzegovinian tombstones – who made them and why*, Südost-Forschungen 21, München 1962, 102-143.

Proceedings of the XII Serbian-Bulgarian Astronomical Conference (XII SBAC) Sokobanja, Serbia, September 25-29, 2020
Editors: L. Č. Popović, V. A. Srećković, M. S. Dimitrijević and A. Kovačević
Publ. Astron. Soc. "Rudjer Bošković" No 20, 2020, 155-167

ВЕНЕРА У МИТОЛОГИЈИ ЈУЖНИХ СЛОВЕНА

АЛЕКСАНДРА БАЈИЋ

Друштво за археоастрономска и етноастрономска истраживања „Влашићи", Београд
E-mail: aleksandra.bajic@gmail.com

Резиме. У раду је тестирана хипотеза да народно литерарно стваралаштво балканских Словена може да представља добар и поуздан извор за изучавање њихове предхришћанске религије и митологије. Како религија и митологија по правилу укључују и схватање света у одређеној култури т.ј. космологију, у истим изворима се могу потражити и знања и схватања о небу и небеским телима и начин на који су та схватања уграђена у митове и обреде. У конкретном случају, испитано је место планете Венере у предхришћанској религији балканских Словена.

УВОД – митологија, космологија и женски принцип

Из најдавнијих праисторијских времена палеолита, наследили смо стеатопигичне статуете, данас популарно назване Венерама. Облине њихових камених или коштаних тела су доведене у везу са плодношћу и прастаром људском чежњом за обиљем хране и топлине. Одатле директан асоцијациони низ води до представе МАЈКЕ, како оне земаљске, „моје мајке", тако и оне космичке, „Велике Мајке", која је родила и однеговала све што постоји. Тако схваћена, мајка постаје један архетип, са више међусобно повезаних значења.

Представа „моја мајка" стоји у корену свих осталих представа „Велике мајке". Њен значај није потребно даље објашњавати јер свако на њу може да пројектује доживљај своје сопствене мајке и утка своје, појединачно искуство, стечено током одрастања. Збир тих појединачних представа ће имати много заједничких елемената, који ће обликовати представу „Велике мајке".

У палеолиту, Месец је сматран регулатором земаљских процеса: време је мерено према његовим фазама, које данас називамо синодичким Месечевим циклусом. Фаза пуног Месеца је сматрана предусловом за плодност, фаза тамног Месеца је била време сексуалног уздржавања, дакле неповољна за активности чији је циљ плодност (Knight, Cris, 1987). Не може се доказати, али изгледа сасвим могуће да су праисторијске фигурине, данас назване Венерама, заправо представе Велике Мајке у лику Месеца (Луна), највероватније оног пуног, имајући у виду заокружено тело. Тако би Велика Мајка регулисала понашање будућих земаљских мајки (и очева).

Са доласком неолита (са земљорадњом и сточарством), представа Велике Мајке се обогаћује: она постаје Велика Мајка Земља, која у хијерогамији са Сунцем зачиње и рађа плодове земље, биљке, животиње и људе, дајући обиље. Али, за плодност земље, Сунце није довољно: неопходна је вода, која може доћи са Земље, као вода извора, потока и река, али и са неба, као атмосферска вода, киша. Како се старија веровања тешко потпуно бришу из свести људи, Велика Мајка у овом тренутку има свој лик Велике Мајке Земље, али и Велике Мајке Месец (којој се додељује власт над водама).

Најразвијеније културе бронзаног доба налазе начина да изградњом монументалних иригационих система, састављених из канала и акумулација, загосподаре токовима вода, које ће ставити у службу земљорадње и сточарства. Готово да није потребно посебно помињати обимне хидрорегулационе системе пронађене у градовима културе долине Инда или оне у Месопотамији. Културе бронзаног доба су јасно патрицентричне: Месец у некима од њих губи своју женску природу, па у Месопотамији постаје Бог, по имену Намму-Суен. Тако, Велика Мајка Земља добија новог партнера, божанство олује и грома, у чијој је компетенцији, уз небеску ватру (гром) и небеска вода, киша. Компетенција над плодношћу припада и новој Богињи, Инани (или Иштар), којој се приписује власт над сексуалним односима, који су предуслов сваке плодности. Нова Богиња је персонификована планета Венера, чије се „понашање“ на небу помно посматра и бележи. Велика Мајка добија још једну одредницу: седам сумерских таблица о Постанку света именују Тиамат, Богињу Хаоса, као примордијалну Велику Мајку, од које ће настати све што постоји.

У старом Египту, уочавају се неке друге промене: Нут (небо) постаје Богиња док Шеб (Земља) постаје Бог.

Укратко, Велика Богиња Мајка је замишљена као:

1. ХАОС (као Тиамат)
2. МЕСЕЦ (Као Артемида или Дијана или Потниа Терон-Мајка животиња)
3. ЗЕМЉА (као Геа, мајка Богова и Титана или као Кибела, мајка Богова и градова)
4. ВЕНЕРА (као Инана или Иштар или Астарте)

На први поглед, чини се да Хаос (Тиамат) некако не спада у овај скуп (или логички низ), јер су сви остали чланови скупа небеска тела. Ипак, Хаос, као супротност организованом Космосу, схваћен као прародитељка Космоса, заправо га употпуњује.

Човек не може да функционише у хаосу. Потребан му је какав-такав ред, који може да влада само у уређеном Космосу, који је рођен из Хаоса. Ред омогућава да човек преживи, развије своје потенцијале и контролише свој страх. Потребно му је схватање света, космологија, да би направио ред у својим мислима и осећањима.

Рекосмо да су сви остали аспекти Велике Богиње Мајке, осим Хаоса, персонификована НЕБЕСКА ТЕЛА[1]. Тиме смо дошли до првих посматрања неба и небеских тела, као најупадљивијег извора правилности и реда. Данас се може доказивати да је посматрање неба започето још у време палеолита, када су већ биле познате Месечеве фазе, које чине циклус, који се данас назива синодичким, и управљале многим аспектима живота људи.Већ неолит доноси знања о дневном и годишњем привидном кретању Сунца (које постаје Бог), заједно са детаљнијим знањем о привидном кретању Месеца – знање о његовим хоризонтским границама, које се понављају у циклусу од 18.6 година, у тесној вези са циклусима у току којих се јављају помрачења Сунца и Месеца. Убрзо и Месец, који је до тада био Богиња, постаје Бог. Бронзано доба доноси знања о привидном кретању планете Венере, која је најсјајније небеско тело после Сунца и Месеца.

ВЕНЕРА У МИТОЛОГИЈИ ЈУЖНИХ СЛОВЕНА

Наслов овог поглавља звучи готово јеретично. Увређено је схватање да је предхришћанска митологија Јужних Словена недовољно истражена а њени извори недовољно сачувани и контрадикторни. Балкански Словени су покрштени између 9. и 12. века, најпре племство, потом народ. Али, народ је конзервативан, спреман да прими ново, али и да чува старо. Чести ратови чине да се људи са сетом сећају старих времена и Богова. У њихово хришћанство уграђени су многи елементи старе религије, сачувани до данас.

Старија литература о народном знању о небеским телима се своди на пар књига и неколико етнолошких и студентских истраживачких радова. Најеминентнији српски истраживачи фолклора и митологије Јужних Словена, Веселин Чајкановић и Сретен Петровић, нису се много бавили народном космологијом. У последњих неколико година, настају значајни радови Ђоре Ценева и Никоса Чаусидиса из Македоније и Витомира Белаја из Хрватске, али се ни један од њих није бавио народним знањима о планети Венери. Ипак, постоји велики корпус народног литерарног стваралаштва, у главном забележен у 19. веку, у ком се и данас могу тражити трагови народних знања и веровања о небу. Постоји и обимна етнографско-етнолошка литература, која, истини за вољу, није центрирана на народну предхришћанску космологију, али која ипак садржи извесну количину података који се односе на ту тему.

[1] Овде треба да нагласимо да постоје и веома важне Богиње, које нису персонификације небеских тела.

Сва ова доступна литература, колико је ауторки овог рада познато, осим мањих фрагмената, није преведена ни на један од светских језика, па је недоступна истраживачима изван Балкана.

Обим овог рада не дозвољава ширу анализу расположивог материјала: биће коментарисано поглавље из књиге Ненада Ђ. Јанковића, астронома, о Венери у народној религији Срба и прегледано неколико народних песмама, српских, хрватских и бугарских, које се односе на планету Венеру. Укупан број таквих песама није много већи од броја оних, које су поменуте у овом раду.

Ненад Ђ. Јанковић (Јанковић, 1951, стр. 117-124) наводи да су за Венеру постојала многа народна имена: када планета излази пре Сунца, називали су је Даница и то је њен најчешћи али не и једини назив. Звали су је и Зорњача. Према Дарини Младеновој (2006, стр. 52-53), код Бугара је називана Деница, Зора или Зорница, јер се појављује на истоку у зору, пре Сунца, да би, након његовог изласка, избледела на дневној светлости. Када се Венера појављивала увече, на западу након заласка Сунца, Срби и Хрвати су је звали Вечерњача или Сјајница или Овчарица. Ово последње је, према Ненаду Ђ. Јанковићу, зато што је опомињала овчаре да би, у време њеног појављивања на небу, морали да затворе своје овце у торове, како би биле безбедне. Бугари су је називали Големата „звезда" или Вечерната „звезда".

Јанковић се пита да ли су балкански Словени знали да су Даница и Вечерњача једно исто небеско тело. Налази одговор на то питање у још једном имену за Венеру, а то је Преодница, јер прелази (преоди) са једне стране неба на другу (са истока на запад и обрнуто) а да се никада не могу видети и њен излазак и залазак у току истог дана. Ово зато што је њена максимална угаона удаљеност од Сунца (елонгација) око 47°, чешће се може видети на значајно мањој удаљености, све до минималне од око 15°. Када је ова угаона дистанца још мања, не може се видети уопште, јер је небо сувише светло. Аутор сматра да ово име сведочи да је некада међу балканским Словенима постојало знање да су Даница и Вечерњача један исти небески објекат, али да то знање у народним песмама није чврсто фиксирано јер су их, у време када су песме записане, све чешће преносили (обредни) певачи са недовољним астрономским знањем. Тада су већ увелико постојали црквени и цивилни календари. Народ више није морао да се ослања на сопствена знања о небу и небеским телима како би мерио проток времена, као што је то морао у време турске окупације.

Венера је увек у близини Сунца, те би се очекивало да то буде и у митовима, као Сунчева супруга или сестра.

Однос Венере са Месецом је нешто компликованији: У време када излази Даница (на истоку) у њеној близини може бити само Месец у фази опадајућег српа. Пун Месец ће бити близу свог заласка, на западу. Тако, када народна песма каже: „Звезда тера Месеца, за гору га затера", то може да означава баш ову ситуацију, јер појава Данице „тера" пун Месец са неба. Симблично, Даница и пун Месец су „у свађи". Јанковић налази песму, која описује овакву свађу у Петрановићевој збирци (*Женске народне пјесме*, књ. 1, песма бр. 5).

Даница је карала Мјесеца:
„Ђе си био, мој сјајан Мјесече,
Где си био, где си дангубио?...

Ови стихови могу да имају и другачије тумачење: астрономски гледано, веома танак (млад) растући срп („млад Месец“) претходне ноћи није био видљив на небу, па Даница из песме има право да се забрине. Али, ово може бити под сумњом, јер су Даница и млад Месец увек на супротној страни неба – Даница на истоку а млад Месец на западу.

У време када се појављује Вечерњача (на западу) у њеној близини може да буде растући, „млад“ Месечев срп (пре прве четврти). Тако, Вечерњача може симболично да буде „сестрица младог Месеца“.

Исто тако, „сестрица“ може и да жени свога „младог“ брата, што је забележено у две песме из збирке Вука Караџића (*Женске народне пјесме*, књига 1, песме бр. 230 и 231):

Радује се звијезда Даница:
„Ожeнићу сјајнога Мјесеца,
Испросићу Муњу од облака....“

Ове привидне противречности у песмама могу да буду помирене, ако и народни песник и његова публика знају да су Даница и Вечерњача једно исто небеско тело.

Одавно су истраживачи словенске митологије уочили да у народном литерарном стваралаштву јужних Словена постоје "елементи мита". Ти "елементи" понекад нису ситне честице, на против, у неким песмама су се очували значајни делови паганских митова. Сасвим ретко, поменуто је и име предхришћанског божанства, па и неки његови атрибути. Таква је бугарска народна песма (*Българско народно творчество*, 1963, књига 6. стр.103):

ПЕСМА ВИДУ

Засвирала два овчара:
"Виде[2] ле,

[2] Вид је балканско име Световида (или Свантевита или Свантевида), врховног божанства Балтичких Словена. О том божанству постоје исцрпни историјски извори (Сахон Граматикус, *Дела Данаца* (*Gesta Danorum*) и Хелмхолд фон Бозау, *Словенска хроника* (*Chronica Sclavorum*). Коришћен је превод делова ових текстова на Српски језик Растка Костића, објављен у његовој књизи *Пад Арконе и сумрак словенског паганизма)* Име Вид је ретко директно поменуто у народним песмама балканских Словена, мада га ипак има. У једној се појављује под именом Вид Жеравица, у другој као Вид Даничић, у оба случаја јасно асоцира на Сунце. Боље се сачувало у топономастици и ономастици, о чему је опширно писао Сретен Петровић, један од најзначајнијих савремених истраживача предхришћанске митологије Срба.

Момни ле,
Кулади ле!"
Два овчара, два другара,
Крај Видова равна двора,
С два кавала[3], два јудурма[4],
Да изађе лепи Виде.
Не изађе лепи Виде,
Већ његова невестица,
Па погледа горе, доле,
Она виде два овчара,
Где свирају два кавала,
Два кавала два јудурма,
Па се назад поврнула:
"Ајд' изађи, калин' Виде,
Да ти видиш два овчара,
Два овчара, два другара,
Где свирају два кавала,
Два кавала, два јудурма,
Крај тих наших белих двора!"
Тад изађе лепи Виде:
Пулије му ситне звезде,
На грудима – јасно Сунце,
На плећима јасни Месец.
Пак огреја земљу, небо,
Земљу, небо, гору, воду,
Три планине, сива стада,
Овчаре на ливадама,
Ергене на чардацима,
невестице на чесмама,
Младе моме с погачама,
Старе бабе код огњишта,
А и старце по крчмама.
Он наздрави свакој кући:
"Од Бога вам добро здравље."

[3] Кавал је народни дувачки музички инструмент, свиран на истоку Србије и у Бугарској, у сточарским крајевима.У Бугарској има и обредну улогу.

[4] Реч „јудурм" је турцизам, значење би се могло дати само описно, као „добро усаглашено" или „добро наштимовано" или „сазвучно" или „у садејству". Реч (данас) има и негативну конотацију, када се користи онда када група људи синхронизовано и организовано обавља неку криминалну делатност, на пример крађу. Управо зато што је реч стара, страна и изишла из употребе, остављена је у преводу таква каква јесте. Објашњење речи добили смо од Светлозара Рулинског, сарадника из Бугарске.

Други стих нам јасно каже да је ово обредна, коледарска (кукерска) песма, јер напев садржи „куладе, ле". Да је песма српска, тај напев би гласио „коледо!" или „Ој коледо!" Коледарске песме се певају око Божића, када дани полако почињу да бивају све дужи, након краткодневице. Атрибут "калин" (Виде) асоцира на женску обредну поворку калинарки[5] , која је у исто време пролазила селима и певала своје, "калинарске" песме. Исти атрибут има и симболично значење "младожења", јер је калина (лат. Ligustrum vulgare) обредна сватовска биљка, како код јужних тако и код источних Словена (Руса[6]) .

Последњи стих помиње Бога, али не каже тачноког. Зато, ова песма може да буде један религијски синкретизам, покушај мирења старе словенске вере са новом а може да буде и сасвим чист остатак старе вере, јер је Вид јасно именован а његово "наздрављање" може да представља и његов сопствени благослов свакој кући. Песма јасно каже да Вид ГРЕЈЕ, што је први директан доказ његове соларне природе, уз „јасно Сунце" на грудима.

Вид је у очигледној вези са небом: Сунце је на његовим грудима, звезде су пулије (украсна дугмад) на његовом оделу, Месец му је на плећима. Видов ИЗЛАЗАК греје народ, стада, гору, воду, земљу, небо… Шта би, осим Сунца, могло све то да греје? Песма чак помиње ТРИ ПЛАНИНЕ, што нас подсећа на три брда[7] , која су често постојала поред словенских светих места, као и на бројне планине на Балкану које се зову Триглав или Троглав. Свирка двојице пастира са почетка песме звучи као део обреда или молитва за излазак соларног божанства. Али, на њихову молитву, Сунце (Вид) не излази одмах, најпре излази његова невеста. Ко би могла да буде та невеста? Ко (или шта) излази на истоку пре Сунца? Наравно, то је "звезда" Даница, Деница или Зорница, како народ на Балкану назива планету Венеру. Да би смо били сасвим сигурни, прочитаћемо једну српску народну песму, из збирке Јаше Продановића (1925, *Женске народне пјесме*, песма бр. 314) :

Девојка је Сунцу говорила:
"Јарко Сунце, љепша сам од тебе!
Ако ли се томе не верујеш,

[5] Обичај је до скора био сачуван код Срба на Косову и Метохији. Обредну поворку су чиниле девојке, од којих је једна била маскирана као невеста, а две као старци, са дугим белим брадама. Остале учеснице обреда су биле без маски. (Кулишић, Ш; Петровић, Петар, Ж; Пантелић, Никола: (1998) *Српски митолошки речник*, Етнографски институт САНУ, Београд, стр. 235).

[6] *Толковьи словарь живого великорусского языка* Владимира Дальа: (1883-1886), Общества любителей российской словесности, Москва (Том 2, калина).

[7] Биографи Ота Бамбершког, Ебо, Херборд и неименовани монах из манастира Прифлингер, описују град Волин, који се налазио у подножју ТРИ БРДА, на највишем од тих брда се налазио храм Триглава, словенског божанства са три главе. Три брда код нордијског светилишта у Упсали, описује Сакс Граматик, а могу се видети и данас. Простор светилишта је и данас у облику круга.

Ти изађи на то равно небо,
Ја ћу изаћ' у гору, на воду!"
Кад је јутро ведро освануло,
Излазило на небо Сунашце,
А ђевојка за гору на воду.
Угледа је лијепо Сунашце,
Угледа је кроз јелово грање,
Колик' се је ашик учинило,
Трипут се је Сунце заиграло,
Па одвуче лијепу ђевојку,
Да је узме себи за љубовцу,
Од ње поста звијезда Даница.

Ово није једина песма у којој се тврди да је соларно божанство ожењено "звијездом" Даницом. Постоји и песма из Далмације (Хрватске народне пјесме, Књ. Vuk St. Karadžić, *Život i običaji naroda srpskoga*[8] (1867) Dostupno na internetu, песма бр. 327), у којој се можда појављује и име Перуна (Пере), словенског Бога грома:

Јидрило дриво низ море,
На њему ПЕРЕ војвода;
На Перу танка кошуља,
Тања од листа макова,
Гушћа од листа наранче.
Мајка ми Перу говори:
„Одкле ти, Пере, кошуља?"
А Пере њојзи казује:
„Када се СУНЦЕ женило
Пресвитлом звиздом Даницом,
Ја сам јој био за кума.
Ја сам јој прстен дарива'
У ком су Сунце и Мисец,
А она мени дарива
Танену ову кошуљу."

Ето, сада нам се чини као да смо решили некакав систем једначина са две непознате (вредности). Схватили смо да је Вид (Световид) исто што и Сунце т.ј. соларно божанство, а да је његова митска супруга персонификована „звезда" Даница. Прва песма нам је јасно показала прву тврдњу (Вид = Сунце) и указала нам на постојање његове митске супруге, а онда су нам друге две песме откриле њен идентитет (супруга Вида или Световида = „звезда" Даница). „Понашање" те супруге је астрономски коректно. Да ли су

[8] Vuk St. Karadžić, *Život i običaji naroda srpskoga* (1867) Dostupno na internet: https://sr.wikisource.org/sr-ec

ове песме (крупни) делови основног мита о небеској природи соларног Бога и његове супруге?

Знамо да су људи веома давно препознали Сунце као извор живота. Многи древни народи су имали своје соларне Богове. По томе, Словени не би били никакав изузетак.

Намећу се и нова питања: какве су се песме певале ако, у време Божића, Венера није била видљива као Даница, већ као Вечерњача? Какве су се певале ако није била видљива уопште? На крају, да ли је могуће да четири лика Световида (Вида) имају само једну супругу? Наравно, није могуће истражити све одједном, доћи ће и та питања на ред.

Да је тачно и оно што су рекли католички мисионари у Померанији и Полабљу, да је Вид (Световид) Бог рата, уверава нас још једна српска коледарска песма, из збирке Вука Караџића (1867, *Живот и обичаји народа српскога*, стр. 7 и 8), која директно помиње Вида и његову невесту:

Војевао бели Виде, Коледо,
Три године с клети Турци,
А четири с црни Угри.
Када Виде с војске дође,
Седе Виде да вечера.
Стаде громот, стаде тропот,
Око двора Видојева.
„Изиђ' љубо те погледај,
Шта је громот, шта је тропот,
Око двора Видојева!“
Кад изађе верна љуба,
Коњи му се копитају,
Радују се господару,
Што је скоро с војске дош'о.
И голуби с крил'ма бију,
Радују се господару,
Што је скоро с војске дош'о.

И у овој песми, која је коледарска као и она бугарска, коју смо мало пре прочитали, Видова љуба (Венера као „звезда“ Вечерњача), појављује се да погледа шта се дешава. Вид оправдано неће изаћи, јер је управо сео (зашао) и вечера – цео догађај се одиграва увече.

Вид је ратник, три године ратује са једним непријатељем, четири са другим. Овим бројевима је можда наговештено тројство али и кватернитет. Још је Питагора знао да тројство симболише хармонију а кватернитет целовитост т.ј. Космос.

Коњи су, према Саксу Граматику, уобичајени атрибут Световида (Вида) и због тога су поменути у овој песми. Симболика коња је тесно повезана са Сунцем још од бронзаног доба. О томе постоје веома убедљиви научни радови. Голубови (као и све птице) симболишу душе, како душе живих, тако и душе предака.

Тако смо добили и податак да је Вид у јасној вези са Сунцем, да је божанство рата, да је „господар душа“, да је истовремено троједан и четворојединствен. Венера је, наравно, у близини.

Ова песма даје за право Ненаду Јанковићу: изгледа да су балкански Словени стварно знали да су Даница и Вечерњача једно исто небеско тело: У бугарској песми о Виду, коју смо прву прочитали, Сунчева невеста је јасно Даница. У овој последњој, она је персонификована Вечерњача.

СИМБОЛИКА И ДИВИНАЦИЈА

Ако је „звезда“ Даница супруга врховног божанства, очекујемо да су се и супруге земаљских владара симболично идентификовале са њом. Песма из збирке Јаше Продановића, забележена у Славонији (Продановић, 1925, песма бр. 335, стр. 239), говори управо о томе:

Санак снила царица госпођа,
да се ведро небо проломило,
сјајан Мјесец на земљицу пао,
а звјездице крају прибегнуле,
сам' Даница остала самица.
Што је снила, све је погодила:
Што се ведро небо проломило,
то су цару разрушени двори;
што је Месец на земљицу пао,
то царева погубљена глава;
што звездице крају прибегнуле,
то цареве остале сироте;
што Даница остала самица,
то царица била удовица.

Овде ћемо се свакако сетити да је Венера у Месопотамији називана „светла краљица неба“ и била персонификована као богиња Инана. И тамо су постојале соларне и лунарне династије овоземаљских владара. За оне прве, помрачење Сунца је сматрано лошим предзнаком по династију, за оне друге, то је било помрачење Месеца.

МОЛИТВА

Ако је планета Венера код балканских Словена била персонифована као божанство, очекивало би се да јој се упућују молитве. Следећа песма из збирке Вука Караџића (Караџић, 1881, репринт 1989, књ. 1, песма бр. 224) говори управо о томе:

Рано рани ђевојчица
и Даницу вјерно моли:
„О Данице, о сестрице,
подај мени свјетлост твоју,
да наресим младост моју...

Молитва ове девојке ће бити услишена у наставку песме.

Постоји слична бугарска народна песма *(Българско народно творчество*, књига 2, стр. 447):

Дена из двора иђаше
и ситне сузе ронише
а према небу гледаше.
На небу сија Зорница.
Дена Зорници казује:
„Зорнице, мила сестрице,
ти тако рано излазиш,
и свашта јасно погледаш,
и горе, сестро, и доле.
Да ли си видела, Зорнице,
мог љубљенога Николу,
Николу, младог војводу?...

У даљем току песме, Зорница ће јој одговорити и потанко испричати све што зна о реченом младом војводи. Дена или Деница је име девојке али и друго бугарско име за Зорницу (Даницу).

Постоји и трећа, слична песма, забележена у часопису *Босанска вила* (1886. XXIV, стр. 381):

О звјездице, моја Преоднице,
Преодила с' небо и земљицу:
Јеси л' била до Будима града?
Јес' видила будимског Стевана?
Спрема ли се Стево у сватове?...

У све три песме, „звезда“ Даница (Зорница) или Преодница је јасно персонификована и могу јој се упућивати молитве, на које се очекује одговор. Није нађена ни једна песма где се Даници (Зорници) или Преодници обраћа мушка особа, што указује на могућност да је код балканских Словена она била схваћена као заштитница жена.

Постоје и неколико песмама, које једноставно казују у које се доба ноћи одвија радња песме. Једна од таквих је записана у Хрватској (*Хрватске народне пјесме,* Књига 1, песма бр. 53)

Јоште није зора забјелила,
Ни Даница лишцем промољела,
Ни Влашићи небо припјешили.
Птица пјевац крилом ударила:
Прилипу се отворише врата...

Ове песме доказују да је небо често и пажљиво посматрано.

ЗАКЉУЧАК

Извори за проучавање митологије Јужних Словена очигледно постоје. У оквиру тих извора могу се тражити (и налазити) и знања и веровања о небу и небеским телима и устројству света. Треба их поново прегледати у светлу нових сазнања археоастрономије и етноастрономије. Уз пуну свест о језичкој баријери, не можемо да се надамо помоћи страних истраживача – већина извора им није доступна. Све четири песме, које су анализиране целе, сагласно виде планету Венеру као супругу врховног соларног божанства, последње четири указују на веровање да она шаље пророчанске снове и одговара на молитве.

Ненад Ђ. Јанковић је приметио да постоје и песме које Даницу дефинишу као „Сунчеву сестрицу, Месечеву првобратучеду“. Оваква митолошка одредница никако не искључује могућност да је она и Сунчева невеста, јер у митологији није реткост да се врховно божанство жени својом сестром (грчки пар, Зевс и Хера, као и египатски Шеб и Нут, су истовремено и брат и сестра и брачни пар).

Пошто словенско соларно божанство, Вид, има четири главе, т.ј. четири своја аспекта или еманације, могуће су и друге божанске партнерке.

Астрономски односи планете Венере са другим небеским телима се мењају у току времена, чиме често могу да се објасне и одређене „контрадикције“ између тврдњи у различитим песмама.

Може се испитати и када се у току године дешава божанска свадба (или хијерогамија) јер се многе народне песме балканских Словена баве свадбама различитих јунака, нарочито оне епске, али и многе лирске, записана је велика група свадбарских песама. То је свакако тема за неко будуће истраживање.

Литература

Примарни извори:

Босанска вила, часопис за књижевност: 1886, Сарајево, бр. XXIV децембар.

Българско народно творчество: 1962, Български писател, София, (Књига 2 и 6).

Женске народне песме, антологија: 1925, приредио Јаша Продановић, Издавачка књижарница Геце Кона, Београд.

Караџић Вук Ст.: 1841, *Српске народне пјесме*, репринт Нолит, Београд, (1989) (Књига 1, Женске народне пјесме)

Караџић Вук Ст.: 1867, *Живот и обичаји народа српскога,* Доступно на интернету. хттпс://ср.wикисоурце.орг/ср-ец.

Петрановић Богољуб: репринт 1989, *Српске народне пјесме из БиХ*, Свјетлост, Сарајево (*Женске народне пјесме*, књ. 1).

Хрватске народне пјесме: 1909, Уредио Никола Андрић, Матица Хрватска, Загреб, (Књига I).

Секундарни извори:

Владимира Даль: 1883-1886, *Толковый словарь живого великорусского языка*, Общества любителей российской словесности, Москва.

Јанковић Ненад Ђ: 1951, *Астрономија у предањима, умотворинама и обичајима Срба*, Одељење за друштвене науке САНУ, Београд.

Knight Chris: 1987, *Blood Relations – Menstruation and the Origins of Culture*, Thesis submitted for the degree of Doctor of Philosophy University College London; available online:
http://radicalanthropologygroup.org/sites/default/files/pdf/pub_chris_thesis.pdf

Костић Растко: 2009, *Пад Арконе и сумрак словенског паганизма*, Атос, Београд (превод на српски језик поглавља која се односе на Словене која су написали: Сахон Граматикус: *Дела Данаца* (*Gesta Danorum*) и Хелмхолд фон Бозау, *Словенска хроника* (*Chronica Sclavorum*)

Кулишић Шпиро, Петровић Петар Ж, Пантелић Никола: 1998, *Српски митолошки речник*, Етнографски институт САНУ, Београд

Младенова Дарина: 2006, *Звездното небе над нас – етнолингвистично изследване на балканските народни астроними*, Академично издателство Проф Марин Дринов, София.

Петровић Сретен: 2004, *Српска митологија у веровању, обичају и ритуалу*, Народна књига и Алфа Невен, Београд.

VENUS IN THE MYTHOLOGY OF THE SOUTHERN SLAVES

The paper tests the hypothesis that the folk literary creation of the Balkan Slavs can be a good and reliable source for studying of their pre - Christian religion and mythology. Since religion and mythology, as a rule, include and the understanding of the world in a certain culture, i.e. cosmology, in the same sources one can look for knowledge and perceptions about the sky and celestial bodies and how these perceptions are embedded in myths and rituals. In this particular case, the place of the planet Venus in the pre-Christian religion of the Balkan Slavs was examined.

Proceedings of the XII Serbian-Bulgarian Astronomical Conference
(XII SBAC) Sokobanja, Serbia, September 25-29, 2020
Editors: L. Č. Popović, V. A. Srećković, M. S. Dimitrijević and A. Kovačević
Publ. Astron. Soc. "Rudjer Bošković" No 20, 2020, 169-180

СУНЧАНИ ВАЛЦЕР

ПЕТАР В. ВУЦА

Кикинда
E-mail: vboba@open.telekom.rs

Апстракт . Сунчани часовник је часовник који мери време према положају сунца. Положај сунца се мења, онда се мења и време које сенка показује. Сунчани сатови могу да се прилагоде свакој површини на коју фиксирани објект баца сенку. Сунчани сатови показују само дневно соларно време.Сунчане сатове су познавали у Египту. Развиле су их и друге културе: Кинези, Антички Грци и Римљани. Врста сунчаног сата с показивечем (гномон) описана је у Старом Завету (Исаија 38:2). Сматра се да је математичар и астроном Теодосије из Битиније (око 160. п. н. е. – око 100. п. н. е.) изумео универзални сунчани сат који је могао да се користи било где на свету. У овом прилогу даћемо преглед где се налазе сунчани сатови у Војводини. Ко су аутори сатова ? Из кога времена потичу. У раду ће бити приказане фотографије суначаних сатова: Сомбор,Зрењанин, Крушедол, Панчево, Кумане, Мокрин, Сремској Митровици,Кикинда. Дати податке о сатовима и њиховим ауторима, за које се зна да су аутори.

1. УВОД

Питали паора (чобанина): "Имаш ли сат?"
„Шта ће ми сат кад ја имам сунце". - Одговори он.

" Знајте да морате служити не времену, него богу "
Св. Атанасије Велики

"Часови пролазе и оптужују нас "
Оксфорд

Човек је од најстаријих времена покушавао да омеђи, премери и одмери време, желећи да контролише и омеђи своју коначност. Штап пободен вертикално у земљу, чија сенка је указивала доба дана - *гномон*, најстарији је инструмент за мерење времена. Гномон је прерастао у сунчани сат, који има историјску, уметничку и културну вредност. Сунце је човеков најсавршенији и најстарији сат. *"Ruit hora-час одмиче" говорили су стари Латини.*

Сунчани сатови су познати још од старог Египта, а развиле су их и друге културе: кинеска, старогрчка и римска. Врста сунчаног сата с показивачем описана је у Старом завету. Најстарији писани текст о сунчаним часовницима датира из 732. године п.н.е.

Сматра се да је математичар и астроном Теодосије из Битиније (од 160 - 100. п.н.е.) урадио унивезарлни сунчани сат који је могао да се користи било где на земаљској кугли. Ђузепе Бјанкани објавио је књигу " *Constructio instrumenti ad horologia solaria*" у којој даје упуства за израду савршеног сунчаног сата. Француски астроном Оронс Фине конструисао је 1524. године сунчани сат од слоноваче. Италијански астроном Ђовани Падовани објавио је расправу о сунчаним сатовима 1570, у којој је дао и упутства за припрему и постављање зидних, вертикалних и хоринзонталних сунчаних сатова.

У ведрим летњим данима, Сунце обасјава предмете на Земљи и ствара сенку/е која се правилно креће, упоредо са ротацијом Земље, око своје осе. Чобани су пратили своју сенку да би се оријентисали колико је приближно сати да би стоку терали кући. Осим личне сенке, они су користили и чобански штап. Људи су размишљали како да искористе сунчеве зраке да би направили неки уређај који ће одређивати време. Када је ведро, по сунчаном сату може се одредити приближно време. Постоје разне врсте сунчаних сатова, а најпопуларнији је „Људски сунчани сат" (*Human sundial*). Сунчани сат се израђује у школским двориштима, парковима, јавним површинама, црквама, кућама богатих домаћина јер га је немогуће украсти и уништити. Сунчани сат одређује време према положају Сунца. Положај Сунца се мења, па се мења и време које сенка показује. *Сунчани сат помоћу сенке **игле** показује право време, сунчево време, као одраз стварног положаја Сунца на небу. Сунчани сатови показују само дневно соларно време.*

У животу користимо средње време (зонско), које тече равномерно. За ову интерактивну инсталацију потребна је корисникова сенка за приказ тачног зонског времена.

2. САТ У СРЕМСКОЈ МИТРОВИЦИ

„У пролеће 1981. године откривен је у Сремској Митровици у близини старог православног гробља, фрагментовани римски сунчани сат, израђен као сложена вајарска композиција са три људске фигуре у природној величини. Сунчани сат постављен је на правоугаоној плочи, у лоптасто издубљеној основи. Остала је неоштећена само једна трећина сатног механизма, који носи уобичајено име „скафе". По означеној датумској линији види се да је рађен прецизно за географску ширину Сирмијума. Веома сложена астрономско-математичка анализа показала је да је сат морао бити веома тачан и да је на његову урезану датумску линију сваког 21. марта падала сенка и показивала да је прошло годину дана. Радијалне линије испод тога служиле су за обележавање дневних сунчаних сати." Сунчани сат се свакако налазио на неком богатом маузолеју на западном римском гробљу у

Сирмијуму" То је уједно и најстарији сунчани сат у нашој земљи и има огроман уметнички и научни значај." (Милошевић, 1985).

Слика 1: Сунчани сат у Сремској Митровици. (Фото Ј. Вуца).

„Богати грађанин Сирмијума Кратило Папије предосетио је 100-те године нове ере да му је близу крај живота. Када је диктирао тестамент, постарао се да трајно овековечи породицу Папија, која је у Сирмијуму имала углед и просперитет. За његов маузолеј допреман је бели мермер са острва Брач. Позвао је вајара и астронома-математичара. Вајар је у људској величини израдио статуу Атласа, легендарног вођу Титана у борби против Зевса. После Зевсове победе, Атлас је кажњен да на својим плећима носи читави небески свод. Уместо небеског свода, он на плећима носи сунчани свод" (Милошевић, 1985).

3. ВЕТЕРАН БЕЗ ГРЕШКЕ
Сунчани сат у Сомбору

Сунчани сат се налази на фасади римокатоличког жупног двора и постављен је на видном пољу. Љепота изгледа оваквог сата припада некадашњем професору и управитељу Сомборске учитељске школе Јовану Чокору. Јован је рођен 1810. године у Баји, а умро 1871. године у Сремским Карловцима. Јован Чокор, био је професор мађарског језика, педагогије, методике, угарске историје и артиметике у сомборској Српској учитељској школи (Препарандији). Управник ове школе је од 1847 до 1852.године. Чокор се аматерски бавио астрономијом. Из Сомбора Јован је отишао у октобру 1853. године, у Сремске Карловце. Ту се замонашио у манастиру Кувеждин на Фрушкој гори, узевши монашко име Јулије. Патријах Рајачић га шаље у Беч, да изучи штампарске вештине. ***последњи***” „EGY EZEKBÖL VĚGÖRĂD!“.

Када се вратио у Сремске Карловце, Јован 1858. године, буде именован за управника српске Митрополитско-гимназијске штампарије и ту ради до 1865. године, и буде постављен за архимандрита манастира Грабовац у Барањи. Ко посматра овај сат, уочиће да је између шипке и бројчаника насликан анђео и петао, који кукуриче и најављује зору. Испод бројчаника је крупним словима написано: „***Један ти је од ових***

Овај натпис изазива немир код пролазника. Мирно опомиње на пролазност, неминовност смрти и коначност људског живота.

Слика 2: Лево: Од петла до анђела - сунчани сат у Сомбору; десно: Јован Чокор.

Вељко Петровић српски књижевник, написао је лепе стихове посвећене сунчаном сату, вероватно инспирисан ***сунчаним сатом који је гледао у свом вољенм родном Сомбору.***

Сунчани часовник

Horas non numero nisi serenas

Налете прамен облака на сунце. *И неста сенке, казаљке, на плочи.*
Да л је то време стало? Бар за нас, За поклонике светлости, бистрине?

То што ми људи, немушти, зовемо временом, веком, годином, тренутком,
то зјело вечно, мрачног постојања, што билом својим меримо тануцким,

и магновеним, пролазним усхитом, где су и осе, полови, безбројни, а све и ништа
исто, подједнако, горе што доле, лево што и десно, исток и запад ко север и југ,
све саме речи и обмане нашег, ко воден цвет прекратког битисања.

Док назиремо и слухтимо плахо, одвасуд како притиче и носи незауставног
тог трајања ток, ко разјапљене празне шкољке шум, потмуле плиме,
глуве ноћи хук ...

Стога чекај: Звезда чим опет просија, и сенка стреле, такнуте у жижу,
падне на бројку и на танки зарез. Прени се, неће прнути ни часак, а,
све се пјано у светлости топи!

И цвет, ко дете на образу с капљом, насмешиће се руци што га бере,
и кишна млака, у јарку крај пута,
у зрцало се небеса претвара!...

Вељко Петровић

4. МАНАСТИР У КРУШЕДОЛУ
Бог је господар нашег времена

Слика 3: Сунчани сат у Крушедолу. Фото Т. Филиповић.
Име аутора овога сата нисам пронашао.

5. САТ У ПАНЧЕВУ
Бројим само ведре тренутке твога живота

У граду Панчеву су војне и цивилне зграде пројектовали инжењерски официри. Изградњом је руководио мајор Флајшман. Планска урбанизација Панчева приписује се бригадном генералу Миховилу Михаљевићу, који је градом управљао од 1816. до 1831. године. У његовој кући, која је направљена 1750. године, било је седиште Банатске војне границе, чији је командант био Миховил Михаљевић.

Накнадно је на јужној фасади зграде урађен сунчани сат (***, 1982).

Слика 4: Сунчани сат у Панчеву.

Не постоји тачан податак о години градње сунчаног сата, али се води да је направљен око 1750.године. Званичних података о томе ко је направио сунчан сат, нема. Има само две претпоставке: прва, да је **сунчани** сат направила група занатлија, и друга, да је уметник, који је направио портрет Михаљевића на зиду, направио и сунчани сат.. Једноспратни угао зграде затварају улице Николе Тесле и Петра Драпшина. Фасада има седам прозора на јужној страни и сунчани сат, док је на источној страни. средњи део наглашен рустиком и балконом са оградом од кованог гвожђа (***, 1982).

Слика 5: Лепотан на јужној страни улице - сунчани сат у Панчеву. Фото Р. Ђерић.

Садашњи власник је Душан М. Ћурчин, адвокат. У склопу куће држи апотеку, која носи назив ***„КОД СУНЧАНОГ САТА"*** . Душан М. Ћурчин каже да он није власник целе куће, већ само једног стана на спрату, док је остало државна имовина[3]

6. СУНЧАНИ САТ У ЗРЕЊАНИНУ
Запамти дете, ја означавам време које ти губиш

Сат се налази на згради Економске школе у Зрењанину. Испод њега пише: ***Бројим само сунчане сате.*** Аутор сата је мр Крсте Наумовски, професор, а мурал је урадио мр Милутин Мићић професор. Сат је урађен 1995. године.

Слика 6: Дан настаје попут сенке. - Сунчани сат у Зрењанину. Фото Т. Мун ђан.

7. СУНЧАНИ САТ У КУМАНУ
Сенка теби, Сунце мени.

У Куману, на забату куће Анђе Трифуњагић, ћерке некадашњег власника Лазе Трифуњагића, сеоског берберина, угледне личности села Кумане, налази се сунчани сат. Кућа је направљена 1912. године, а сат вероватно урађен двадесетих година XX века. Ове податке добио сам од др Душка Летића, професора на Техничком факултету „М. Пупин" у Зрењанину. Данас је то Улица Иве Лоле Рибара бр. 20. На сату , који је украшен , може се читати пола сата и цео сат.

Слика 7: Сунчани сат у Куману . Фото. Д. Летић.

8. САТОВИ У КИКИНДИ
Овако пролази слава
Кад Сунце греје ја се смејем

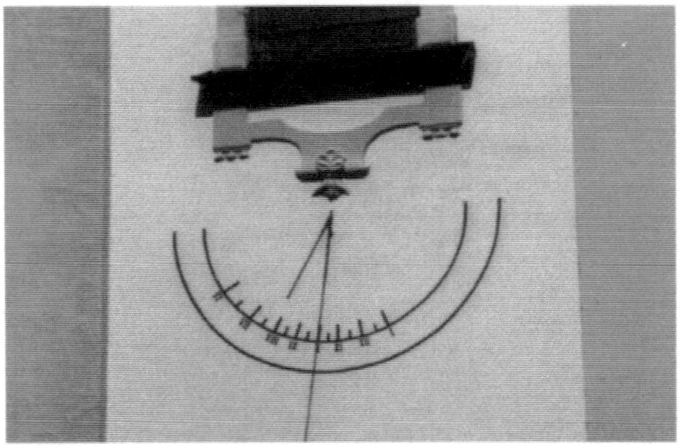

Слика 8: Сунчани сат у Кикинди. Фото В. Сретеновић.

На кикиндској православној цркви налази се стари сунчани сат, рад непознатог аутора.

Слика 9: Сунчани сат у Кикинди Фото. В. Сретеновић.

Сунчани сат који се налази на згради Авала урађен је 2002. године, да би Кикинђане подсећао на дане када је било тотално помрачење Сунца у овом граду 1999. године. Зграда се налази у центру града. Сат су урадили проф. др Милутин Тадић са Географског факултета у Београду и проф. др Петар В.

Вуца са Високе техничке школе струковних студија у Зрењанину. Диск од глине направио је професор Милорад Кнежевић из Башаида.

Враћам,***сунчани сат*** на оно време
Које ништа не откуцава,
И ништа не окончава.

О, то милозвучно, безвремено време,
Што није суза, није сенка, није бреме,

Ђоко Стојчић
Велики прасак
(Дечије новине – Горњи Милановац)

9. СУНЧАНИ САТ НА МОКРИНСКОЈ ЦРКВИ

Храм у Мокрину изграђен је 1762. године и посвећен је светим архангелима Михаилу и Гаврилу. Храм се налази у центру села. На цркви, на источној страни, налази се сунчани сат. Аутор није познат. Сат је урађен 29. септембра 1817. године. За моловање сунчаног сата издвојено је 94 форинте и 58 крајцара (црквена књига рачуна).

Слика 10: Сунчани сат у Мокрину. Фото П. Вуца.

10. СУНЧАНИ САТ У ТИТЕЛУ

Слика 11: Сунчани сат у Тителу. Фото А. Николић.

Сат се налази у центру града. Није видљив, јер се налази у парку. Нема имена аутора нити кад је рађен.

Приликом тражења података о сунчаним сатовима закључио сам да их има највише на црквама, а мали број на приватним кућама. Када су сатови на приватним кућама, онда су то били богати домаћини. Интересантно је да нигде нема забележено име аутора сунчаних сатова, ни година њиховог постављања.

Највише сунчаних сатова има у Банату у Војводини. Зашто!? Мислим зато што су Банаћани били, вредни, тихи и богати и волили сатове који раде тихо и нечујно, а увек су ту.

Литература

***: 1982, *Каталог изложбе "Културно историјски споменици" ,Панчевац-градитељско наслеђе*", Народни музеј у Панчеву.

Милошевић Петар: 1985, Сунчани сат из Сирмијума, *Старинар* XXXVI, Археолошки институт, Београд.

П. В. ВУЦА

SUNDIAL WALTZ

A sundial is a clock that measures time according to the position of the Sun. As the position of the Sun changes, the shadow shows time. Sundials can be adjusted to any surface on which a fixed object throws a shadow. Sundials show only a day's solar time. The sundials were known already in Egypt. They were also developed by other cultures: the Chinese, the Ancient Greeks and the Romans. The type of sundial with a pointer (gnomon) is described in the Old Testament (Isaiah 38:8). It is believed that the mathematician and astronomer Theodosius of Bithynia (about 160 BC - about 100 BC) invented an universal sundial that could be used anywhere in the world. In this report we will give an overview of the location of sundials in Vojvodina. Who are the makers of the clocks? From which period do they origin? The work will show photographs of sundials: Sombor, Zrenjanin, Krušedol, Pančevo, Kumane, Mokrin, Sremska Mitrovica, Kikinda and offer data on the clocks and their makers, those who are known for sure to to be the creators.

LIST OF PARTICIPANTS

Victor Afanasiev
Special Astrophysical Observatory of Russian Academy of Science, Nizhnij Arkhyz, Zelenchukskiy Region, Karachai-Cherkessian, Republic, Russia 369167

Jovan Aleksić
Department of Astronomy, Faculty of Mathematics, University of Belgrade, Studentski trg 16, 11000 Belgrade, Serbia

Bojan Arbutina
Department of Astronomy, Faculty of Mathematics, University of Belgrade, Studentski trg 16, 11000 Belgrade, Serbia

Simeon Asenovski
Space Research and Technology Institute - BAS, Acad. Georgi Bonchev St., bl. 1, Sofia 1113, Bulgaria

Rumen Bachev
Institute of Astronomy and NAO, BAS, 72 Tsarigradsko shosse, Sofia, Bulgaria

Aleksandra Bajić
Society for Archaeoastronomical and Ethnoastronomical Research "Vlašići", Belgrade, Serbia

Svetlana Boeva
Institute of Astronomy and NAO, BAS, 72 Tsarigradsko shosse, Sofia, Bulgaria

Mariyana Bogdanova
Department of Astronomy, Faculty of Physics, University of Sofia, 5 James Bourchier Blvd., Sofia 1164, Bulgaria

Ljube Bojevski
Ss Cyril and Methodius University, Skopje, North Macedonia

Daniela Boneva
Space Research and Technology Institute - BAS, Acad. Georgi Bonchev St., bl.1, Sofia 1113, Bulgaria

Srdjan Bukvić
Faculty of Physics, University of Belgrade, Studentski trg 12-16, Belgrade, Serbia

Magdalena Christova
Department of Applied Physics, Technical University - Sofia, 8 Blvd Kl. Ohridski, 1000 Sofia, Bulgaria

Zorica Cvetković
Astronomical Observatory, Volgina 7, 11060 Belgrade, Serbia

Goran Damljanović
Astronomical Observatory, Volgina 7, 11060 Belgrade, Serbia

Yordan Darakchiev
Department of Astronomy, Sofia University, 5 James Bourchier St., 1164 Sofia, Bulgaria

Milan S. Dimitrijević
Astronomical Observatory, Volgina 7, 11060 Belgrade, Serbia

Branko Dragovich
Institute of Physics Belgrade, Pregrevica 118, Belgrade, Serbia

Katya Georgieva
Space Research and Technology Institute at the Bulgarian Academy of Sciences, Sofia, Bulgaria

Dragana Ilić
Department of Astronomy, Faculty of Mathematics, University of Belgrade, Studentski trg 16, 11000 Belgrade, Serbia

Isidora Jankov
Department of Astronomy, Faculty of Mathematics, University of Belgrade, Studentski trg 16, 11000 Belgrade, Serbia

Miljana Jovanović
Astronomical Observatory, Volgina 7, 11060 Belgrade, Serbia

Daniela Kirilova
Institute of Astronomy and NAO, BAS, 72 Tsarigradsko shosse, Sofia, Bulgaria

Boian Kirov
Space Research and Technology Institute at the Bulgarian Academy of Sciences, Sofia, Bulgaria

Aleksandra Kolarski
Technical Faculty "Mihajlo Pupin", University of Novi Sad, 23000 Zrenjanin, Serbia

Kostadinka Koleva
Space Research and Technology Institute at the Bulgarian Academy of Sciences, Acad. Georgy Bonchev st., bl. 1, 1113 Sofia, Bulgaria

Renada Konstantinova-Antova
Institute of Astronomy and NAO, BAS, 72 Tsarigradsko shosse, Sofia, Bulgaria

Ognyan Kounchev
Institute of Mathematics and Informatics, Bulgarian Academy of Sciences, Acad. G. Bonchev st., bl. 8, 1113 Sofia, Bulgaria

Andjelka Kovacević
Department of Astronomy, Faculty of Mathematics, University of Belgrade, Studentski trg 16, 11000 Belgrade, Serbia

Jelena Kovačević Dojčinović
Astronomical Observatory, Volgina 7, 11060 Belgrade, Serbia

Maša Lakićević
Astronomical Observatory, Volgina 7, 11060 Belgrade, Serbia

Eugene Malygin
Kazan Federal University, Kazan, 420008 Russia

Sladjana Marčeta Mandić
Astronomical Observatory, Volgina 7, 11060 Belgrade, Serbia

Lyubov Marinkova
Department of Astronomy, Faculty of Physics, University of Sofia, 5 James Bourchier, 1164 Sofia, Bulgaria

Boyko Mihov
Institute of Astronomy and NAO, BAS, 72 Tsarigradsko shosse, Sofia, Bulgaria

Evgeny Mikhailov
M.V. Lomonosov Moscow State University, 1 Leninskie gory, Moscow, 119991, Russia

Stanislav Milošević
Department of Astronomy, Faculty of Mathematics, University of Belgrade, Studentski trg 16, 11000 Belgrade, Serbia

Rositsa Miteva
Institute of Astronomy and NAO, BAS, 72 Tsarigradsko shosse, Sofia, Bulgaria

Yoana Nakeva
International Hall, London, UK

Grigor Nikolov
Institute of Astronomy and NAO, BAS, 72 Tsarigradsko shosse, Sofia, Bulgaria

Aleksandra Nina
Institute of Physics, Belgrade, Pregrevica 118, Belgrade, Serbia

Petya Pavlova
TU Sofia, Branch Plovdiv, Plovdiv, 63 "Sankt Petersburg", Bulgaria

Rade Pavlović
Astronomical Observatory, Volgina 7, 11060 Belgrade, Serbia

Nikola Petrov
Institute of Astronomy and NAO, BAS, 72 Tsarigradsko shosse, Sofia, Bulgaria

Jelena Petrović
Astronomical Observatory, Volgina 7, 11060 Belgrade, Serbia

Jovana Petrović
Mathematical Institute SANU, Kneza Mihaila 36, Belgrade, Serbia

Luka Č. Popović
Astronomical Observatory, Volgina 7, 11060 Belgrade, Serbia

Vojislava Protitch - Benishek
Astronomical Observatory, Volgina 7, 11060 Belgrade, Serbia

Djordje Savić
Astronomical Observatory, Volgina 7, 11060 Belgrade, Serbia

Elena Shablovinskaya
Special Astrophysical Observatory of the Russian Academy of Sciences SAO RAS, Nizhnyi Arkhyz

Georgi Simeonov
IMI-BAS Acad. Georgi Bonchev Str., Block 8, 1113 Sofia, Bulgaria

Lyba Slavcheva-Mihova
Institute of Astronomy and NAO, BAS, 72 Tsarigradsko shosse, Sofia, Bulgaria

Vladimir Srećković
Institute of Physics, Belgrade, Pregrevica 118, Belgrade, Serbia

Orlin Stanchev
Department of Astronomy, Faculty of Physics, University of Sofia, 5 James Bourchier Blvd., Sofia 1164, Bulgaria

Milan Stojanović
Astronomical Observatory, Volgina 7, 11060 Belgrade, Serbia

Saša Topić
Department of Astronomy, Faculty of Mathematics, University of Belgrade, Studentski trg 16, Belgrade, Serbia

Tsvetan Tsvetkov
Institute of Astronomy and NAO, BAS, 72 Tsarigradsko shosse, Sofia, Bulgaria

Dejan Urošević
Department of Astronomy, Faculty of Mathematics, University of Belgrade, Studentski trg 16, Belgrade, Serbia

Todor Veltchev
Department of Astronomy, Faculty of Physics, University of Sofia, 5 James Bourchier Blvd., Sofia 1164, Bulgaria

Petar V. Vuca
Primary School "Dr Tihomir Ostojić", 23326 Ostojićevo, Serbia

Nikola Veselinović
Institute of Physics, Belgrade, Pregrevica 118, Belgrade, Serbia

Miroslava Vukcevic
Astronomical Observatory, Volgina 7, 11060 Belgrade, Serbia

Krasimira Yankova
Space Research and Technology Institute, Bulgarian Academy of Sciences, Acad G. Bonchev Str., Bl. 1, 1113 Sofia, Bulgaria

AUTHORS' INDEX

Bachev Rumen 15
Bajić Aleksandra 129, 139, 155
Benišek Vladimir 123
Boeva Svetlana 15
Bogdanova Mariyana 61, 69
Bojevski Ljube 79
Čelebonović Vladan 107
Černok Ana 107
Cvetković Zorica 15, 23
Damljanović Goran 5, 15, 23
Dimitrijević Milan S. 129, 139
Dojčinović Ivan 33
Donkov Sava 51
Gavrilov Milivoj B. 123
Ivanov Emil 79
Janc Natalija 123
Jovanović Miljana D. 15, 23
Kirilova Daniela 39
Kolarski Aleksandra 93, 107
Kovačević Anđelka 107
Kovačević-Dojčinović Jelena 33
Lakićević Maša 33
Latev Georgy 15
Marinkova Lyubov 51
Marinković Bratislav P. 107
Marković Gabrijela 15
Marković Slobodan B. 123
Mijić Zoran 107
Milić Žitnik Ivana 107
Miteva Rositsa 79
Nakeva Yoana 79
Nina Aleksandra 107
Panayotova Mariana 39
Pavlović Rade 15, 23
Petrov Nikola 79
Petrovic Jelena 43
Popov Velimir 79
Popović Luka Č. 33, 85, 107, 123
Protić Benišek Vojislava 123
Radovanović Milan 107
Radović Jelena 107
Srećković Vladimir A. 107
Stanchev Orlin 61, 69
Stojanović Milan 15, 23
Tsvetkov Tsvetan 79
Veltchev Todor V. 51, 61, 69
Veselinović Nikola 107
Vince Oliver 15
Вуца Петар В. 169
Vukcevic Miroslava 85
Zdravković Alena 107
Zlatev Ruslan 79

XII SERBIAN-BULGARIAN ASTRONOMICAL CONFERENCE
PROGRAMME

GREEN. INPERSON

BLUE VIRTUAL

BLACK NO DECISION

Friday, September 25

12:00 – 16:00 Arrival
16:00 – 17:00 Registration (at the reception of the Hotel Moravica)

17:00 – 17:30 Opening ceremony
Chairs: **Ognyan Kounchev and Vladimir Srećković**
17:30 – 19:00 Round table: COLLABORATION BETWEEN SERBIAN AND BULGARIAN ASTRONOMERS - PRESENTATION OF THE EXISTING PROJECTS
20:00 - Welcome cocktail

Saturday, September 26

Chair: **Luka Č. Popović**
09:25 – 09:50 Dejan Urošević: PROPER MOTION OF Cyg LOOP FILAMENTS
09:50 – 10:15 Renada Konstantinova-Antova: MAGNETIC FIELD AND ACTIVITY STUDY IN M GIANT STARS
10:15 – 10:30 Isidora Jankov: APPLICATIONS OF MANIFOLD LEARNING TECHNIQUES TO SPECTRAL PARAMETERS OF QUASARS

10:30 – 11:00 Coffee break

Chair: **Katya Georgieva**
11:00 – 11:25 Rositsa Miteva: A NEW SPACE WEATHER SERVICE IN BULGARIA: THE MULTI-ENERGY PROTON EVENT CATALOG
11:25 – 11:50 Miroslava Vukčević: SOLITONS IN THE IONOSPHERE – ADVANTAGES AND PERSPECTIVE
11:50 – 12:15 Yoana Nakeva: POLARIZATION OF WHITE-LIGHT SOLAR CORONA DURING TOTAL SOLAR ECLIPSES
12:15 – 12:30 Saša Topić APPLICATIONS OF PHOTONICS IN EXOPLANETOLOGY

12:30 – 14:30 Lunch

Chair: **Nikola Petrov**

14:30 – 14:55 Milan S. Dimitrijević: THE MODIFIED SEMIEMPIRICAL METHOD (1980-2020)

14:55 – 15:20 Srđan Bukvić: ASTROPHYSICALLY INTERESTING STARK PARAMETERS MEASURED IN LASER-INDUCED PLASMA

15:20 – 15:35 Magdalena Christova: STARK BROADENING OF Be II SPECTRAL LINES

15:35 – 15:50 Mariyana Bogdanova: STUDY OF THE FRACTAL DIMENSIONS IN THE MOLECULAR CLOUD ROSETTE BY USE OF DENDROGRAM ANALYSES

15:50 – 16:05 Zlatko Majlinger, Milan Dimitrijević: STARK BROADENING OF Co II LINES IN STELLAR ATMOSPHERES

16:05 – 16:35 Coffee break

Chair: **Dejan Urošević**

16:35 – 17:00 Aleksandra Bajić VENUS IN THE MYTHOLOGY OF THE SOUTHERN SLAVS

17:00 – 17:25 Milan S. Dimitrijević MYTHOLOGICAL ORIGIN OF CONSTELLATIONS AND THEIR DESCRIPTION: ARATUS, PSEUDO-ERATOSTHENES, HYGINUS

17:25 – 17:40 Aleksandra Bajić: A PAIR OF MONUMENTAL MEDIEVAL BOSNIAN UMBSTONES (STEĆAKS) FROM DONJA ZGOŠĆA

17:40 – 17:55 Petar V. Vuca: SOLAR WALTZ

Sunday, September 27

Chair: **Vladimir Srećković (EUROPLANET SESSION)**

10:30 – 10:40 EUROPLANET INFORMATION

10:40 – 11:05 Aleksandra Nina: ACTIVITIES OF SERBIAN SCIENTISTS IN EUROPLANET

11:05 – 11:30 Nikola Veselinović: COSMIC RAY FLUX MEASUREMENTS AT BELGRADE COSMIC RAYS STATION DURING SOLAR CYCLE 24

11:30 – 11:55 Ognyan Kounchev, Georgi Simeonov: SOME APPLICATIONS OF WAVELET ANALYSIS TO PHENOMENA IN THE IONOSPHERE

11:55 – 12:20 Ramesh Chandra: CHARACTERISTICS OF SEPs ASSOCIATED WITH SOLAR FLARES

12:30 Lunch and afternoon excursion

Monday, September 28

Chair: **Ognyan Kounchev**

09:55 – 10:20 Goran Damljanović: GAIA DR3 AND SOME RESULTS OF SERBIAN-BULGARIAN COOPERATION

10:20 – 10:35 Milan Stojanović: SERBIAN-BULGARIAN OBSERVATIONS OF GAIA ALERTS (GAIA-FUN-TO) DURING 2019

10:35 – 10:50 Miljana Jovanović: COLOR VARIABILITY OF SOME QUASAR IMPORTANT TO THE ICRF -- GAIA CRF LINK

10:50 – 11:20 Coffee break

Chair: **Milan S. Dimitrijević**

11:20 – 11:45 Viktor Afanasiev: A NEW APPROACH TO MEASURING THE SIZE OF THE DUST SUBLIMATION REGION IN AGNs

11:45 – 12:10 Dragana Ilić, Lyuba Slavcheva-Mihova: THE REVERBERATION MAPPING OF QUASARS IN TOTAL AND POLARIZED LIGHT

12:10 – 12:25 Elena Shablovinskaya STOKES POLARIMETER FOR 1-METER TELESCOPE

12:25 – 12:40 Djordje Savić: BROAD LINE POLARIZATION IN ACTIVE GALACTIC NUCLEI: MODELS AND OBSERVATIONS

12:40 – 12:55 Eugene Malygin: THE FIRST RESULTS OF THE PHOTOMETRIC REVERBERATION PROJECT AT THE 1-M TELESCOPE OF SAO RAS

12:55 – 13:10 Orlin Stanchev: TRACING THE LOCAL MORPHOLOGY OF THE MOLECULAR CLOUD ROSETTE USING MOLECULAR-LINE AND DUST-EMISSION DATA

13:05 – 14:30 Lunch

Chair: Dragana Ilić

14:30 – 14:55 Rade Pavlović: LUCKY IMAGING AT AS VIDOJEVICA:PRESENT STATE AND FUTURE PLANS

14:55 – 15:10 Svetlana Boeva: OSCILLATIONS IN KR Aur AT MINIMUM

15:10 – 15:25 Yordan Darakchiev: DEEP LEARNING FOR CLASSIFICATION OF LONG PERIOD VARIABLE STARS IN THE LOCAL GROUP

15:25 – 15:40 Stanislav Milošević: N-BODY SIMULATIONS OF STELLAR STREAMS, BARS, SHELLS AND RINGS IN SPIRAL GALAXIES

15:40 – 16:00 Coffee break

Chair: **Rumen Bachev**

16:00 – 16:25 Nikola Petrov: SHADOW BANDS AND RELATED ATMOSPHERIC CONDITIONS REGISTERED DURING TOTAL SOLAR ECLIPSES

16:25 – 16:40 Yanko Nikolov: SPECTROPOLARIMETRIC OBSERVATIONS OF THE RECURRENT NOVAE

16:40 -16:55 Lyubov Marinkova: EXTRACTION OF A SECOND POWER-LAW TAIL OF THE (COLUMN-)DENSITY DISTRIBUTION IN STAR-FORMING CLOUDS

Chair **Maša Lakićević**

17:00 – 19:00 **Poster presentation**

20:00 Conference dinner

Tuesday, September 29

Chair: **Viktor Afanasiev (Elena Shablovinskaya)**

10:00 – 10:25 Jelena Petrović EVOLUTION OF MASSIVE BINARY SYSTEMS

10:25 – 10:50 Evgeny Mikhailov: NO-Z MODEL FOR MAGNETIC FIELDS OF ACCRETION DISCS

10:50 – 11:20 Coffee break

Chair: **Andjelka Kovačević**

11:20 – 11:45 Branko Dragovich: ON COSMOLOGY OF NONLOCAL GRAVITY

11:45– 12:10 Sava Donkov and Todor Velchev: DENSITY PROFILE OF A SELF-GRAVITATING POLYTROPIC TURBULENT FLUID IN THE CONTEXT OF MOLECULAR CLOUDS

12:10 – 12:30 Closing the conference

12:30 – 14:00 Lunch

LIST OF POSTERS

P01. Bojan Arbutina ON THE DISTRIBUTION FUNCTION OF PARTICLES AT QUASI-PARALLEL COLLISIONLESS SHOCKS

P02. R. Bachev, A. Strigachev OPTICAL FOLLOW-UP OF TRANSIENT EVENTS FROM BELOGRADCHIK OBSERVATORY IN THE ERA OF THE MULTI-MESSENGER ASTRONOMY

P03. Daniela Boneva MID-CYCLE OBSERVATIONS OF CR Boo AND ESTIMATION OF THE SYSTEM PARAMETERS

P04. Natalie Janc MILUTIN MILANKOVIĆ AND ASSOCIATES IN THE CREATION OF THE "KANON"

P05. Daniela Kirilova INFLATIONARY MODELS, REHEATING AND SCALAR FIELD CONDENSATE BARYOGENESIS

P06. Aleksandra Kolarski STORM ACTIVITY OVER BALKAN REGION DURING MAY 2009

P07. Kostadinka Koleva, Rositsa Miteva PRESENTATION OF THE PROJECT "ERUPTIONS, FLOWS AND WAVES IN SOLAR ATMOSPHERE AND THEIR ROLE IN SPACE WEATHER"

P08. Andjelka Kovačević SUPERMASSIVE BLACK HOLE BINARY CANDIDATE PG 1302-102: OSCILLATIONS AND PERTURBATION IN THE PHOTOMETRIC LIGHT CURVE

P09. Jelena Kovačević Dojčinović THE SPECTRAL PROPERTIES OF THE LARGE SAMPLE OF AGN TYPE 2

P10. Maša Lakićević THE SPECTROSCOPIC CORRELATIONS AND MODEL OF DUSTY HYPERBOLOID WITH A THIN DISK

P11. Sladjana Marčeta Mandić THE INFLUENCE OF THE INTERNAL REDDENING ON ESTIMATION OF BLACK HOLE MASS IN QUASARS

P12. Boyko Mihov DATA MINING: ANALYSIS OF THE HQM AND JOMPQ LIGHT CURVES OF AGNs

P13. Boyko Mihov INTRA-NIGHT MONITORING OF THE BLAZAR 0716+714: RESULTS FROM THE 2011 CAMPAIGN

P14. Rositsa Miteva OCCURRENCE RATE AND CAUSAL RELATIONSHIP OF EXTREME SPACE WEATHER EVENTS

P14. Aleksandra Nina LOWER IONOSPHERE DISTURBANCES: THEIR POSSIBLE RELATIONSHIP WITH EARTHQUAKES, AND INFLUENCE ON SATELLITE SIGNALS

P15. Petya Pavlova STUDY OF THE BRIGHTNESS SPREADING IN SOLAR ERUPTIVE PROMINENCE IMAGES

P16. Luka Č. Popović MICROLENSING OF POLARIZED LIGHT OF THE GRAVITATIONAL LENS J1004+4112

P17. Evgeni Semkov, Nikola Petrov SOME OBSERVATIONS ON EXTRATERESTRIAL SOLAR VARIABILTY AND INFLUENCE OVER ATMOSPHERIC OZONE CONCENTRATION AND SOLAR UV RADIATION FLUXES

P18. Lyuba Slavcheva-Mihova STAR FORMATION IN THE HOST GALAXIES OF RADIO-QUIET QUASARS

P.19. Lyuba Slavcheva-Mihova BLACK HOLE MASSES AND BROAD LINE REGION GEOMETRY OF QUASARS

P20. Vladimir Srećković DESTRUCTION OF SOME MOLECULAR IONS OF ASTROPHYSICAL INTEREST

P21. Vladimir Srećković DISTURBANCES OF THE LOWER IONOSPHERE INDUCED BY SOLAR FLARES DURING TRANSITION PHASE OF 24 SOLAR CYCLE

P22. Vladimir Srećković COOPERATION BETWEEN THE ASTRONOMICAL OBSERVATORY IN BELGRADE AND INSTITUTE OF PHYSICS BELGRADE IN INVESTIGATION OF COLLISIONAL AND RADIATIVE ATOMIC PROCESSES IN ASTROPHYSICS"

P23. Saša Topić ROLE OF DIFFRACION LIMITED PHOTONIC SPECTROGRAPHS IN EXOPLANETOLOGY: CASE STUDY OF HPCFC COUPLED ECHELLE SPECTROGRAPH – MODAL STABILITY, SPECTRAL SENSITIVITY AND NOISE FLOOR FOR DETECION OF SUPER EARTHS

P24. Tsvetan Tsvetkov QUASI-PERIODIC VELOCITY FLUCTUATIONS IN ERUPTIVE PROMINENCES OBSERVED BY AIA/SDO

P25. Krasimira Yankova THE ACCRETION DISK IN THE GALACTIC NUCLEI

Photo on the back cover: Participants of the 12th Serbian-Bulgarian Astronomical Conference held in hotel Moravica, Sokobanja (Serbia) from 25-29 September 2020. It was a blended conference where a small number of participants were present in Sokobanja (shown on third panel of first raw).

First row: Saša Simić (Serbia), Jelena Kovačević Dojčinović (Serbia), Đorđe Savić (Serbia), Sladjana Marčeta-Mandić (Serbia), Dejan Urošević (Serbia), Luka Č. Popović (Serbia), Nikola Petrov (Bulgaria), Ljube Bojevski (North Macedonia), Milan S. Dimitrijević (Serbia), Saša Topić (Serbia), Georgi Simeonov (Bulgaria), Ognyan Kunchev (Bulgaria), Vladimir Srećković (Serbia), Aleksandra Nina (Serbia);

Second row: Aleksandra Kolarski (Serbia), Andjelka Kovačević (Serbia), Danijela Boneva (Bulgaria), Doron Chelouche (Israel);

Third row: Eugene Malygin (Russia), Elena Shablovinskaya (Russia), Isidora Jankov (Serbia), Krasimira Yankova (Bulgaria), Theodor Velchev (Bulgaria), Lyubov Marinkova (Bulgaria);

Fourth row: Maša Lakićević (Serbia), Rade Pavlović (Serbia), Goran Damljanović (Serbia), Miljana Jovanović (Serbia), Zorica Cvetković (Serbia), Milan Stojanović (Serbia), Orlin Stanchev (Bulgaria), Rositsa Miteva (Bulgaria).

CIP - Каталогизација у публикацији
Народна библиотека Србије, Београд

520/524(082)

SERBIAN-Bulgarian Astronomical Conference (12 ; 2020 ; Sokobanja)
Proceedings of the XII Serbian-Bulgarian Astronomical Conference, Sokobanja, Serbia, September 25-29, 2020 / [organizer Astronomical Observatory Belgrade, Serbia] ; eds. Luka Č. Popović ... [et al.]. - Belgrade : Astronomical Society "Rudjer Bošković", 2020 (Beograd : Skripta Internacional). - 192 str. : ilustr. ; 24 cm. - (Публикације Астрономског друштва "Руђер Бошковић" ; sv. 20 = Publications of the Astronomical Society "Rudjer Bošković" ; no. 20)

Tiraž 100. - Napomene i bibliografske reference uz tekst. - Bibliografija uz svaki rad. - Registar.

ISBN 978-86-89035-15-5

а) Астрономија - Зборници б) Астрофизика - Зборници

COBISS.SR-ID 28260361